Sustainable Economic Development

Mohamed Sami Ben Ali · Ewa Lechman
Editors

Sustainable Economic Development

Fostering the United Nations Goals

Springer

Editors
Mohamed Sami Ben Ali
College of Business and Economics
Qatar University
Doha, Qatar

Ewa Lechman
Faculty of Management and Economics
Gdańsk University of Technology
Gdańsk, Poland

ISBN 978-981-97-3766-6 ISBN 978-981-97-3767-3 (eBook)
https://doi.org/10.1007/978-981-97-3767-3

This Springer imprint is published by the registered company Springer Nature Singapore Pte Ltd.
The registered company address is: 152 Beach Road, #21-01/04 Gateway East, Singapore 189721, Singapore

Contents

Chapter 1
Institutional Quality, ICT Infrastructure, Transportation, and Sustainable Development: The Case of Lower Income Countries

Rudra P. Pradhan, Mahendhiran S. Nair, John H. Hall, Sara E. Bennett, and Sahar Bahmani

Abstract Transportation is critical for the sustainable development of countries across the globe. A sound transportation system is important for the movement of goods, services, and people worldwide. In many developed economies, significant investments are channelled to develop efficient transportation systems. Intensifying investments in ICT and institutions of governance are critical to ensure that the transportation system deepens its impact on economic growth. Key challenges faced by many lower income countries (LICs) are lack of resources to increase investments in advanced transportation infrastructure and improving governance of the transportation system. This dampens economic growth in LICs. The present study examines the relationship between ICT infrastructure, transportation system, institutional governance, and economic growth in LICs using the panel VAR model from 2005 to 2022. The empirical analysis shows strong short-run and long-run dynamics between the

R. P. Pradhan (✉)
Vinod Gupta School of Management, Indian Institute of Technology Kharagpur, Kharagpur, WB 721302, India
e-mail: rudrap@vgsom.iitkgp.ernet.in

M. S. Nair
Sunway Institute for Global Strategy & Competitiveness and Sunway Business School, Sunway University, 47500 Selangor, Malaysia
e-mail: mahendhiransn@sunway.edu.my

J. H. Hall
Department of Financial Management, University of Pretoria, Pretoria 0028, Republic of South Africa
e-mail: john.hall@up.ac.za

S. E. Bennett
College of Business, University of Lynchburg, Lynchburg, VA 24501, USA
e-mail: bennett.se@lynchburg.edu

S. Bahmani
College of Business, Lander University, Greenwood, SC 29649, USA
e-mail: sbahmani@lander.edu

M. S. Ben Ali and E. Lechman (eds.), *Sustainable Economic Development*,
https://doi.org/10.1007/978-981-97-3767-3_1

variables; hence, careful curation of co-development policies is needed to enable LICs to transition onto a path of sustained economic growth.

Keywords Institutional quality · Transportation · ICT infrastructure · Economic growth · SDGs · Low-income countries

1.1 Introduction

This study examines the intricate relationship between institutional quality, ICT infrastructure, transportation, and sustainable economic growth in lower income countries (LICs), especially small island and landlocked countries. Over the last decade, the importance of well-developed transportation networks has been underscored through the lens of the UN's Sustainable Development Goals (SDGs), with specific relevance to SDG-8 (*decent work and economic growth*), SDG-9 (*industry, innovation, and economic growth*), and SDG-11 (*sustainable cities and communities*). The UN-SDGs emphasizes the importance of good and affordable transportation systems to make cities and human settlements inclusive, safe, and accessible to all segments of the population. Good transportation systems are important for creating vibrant economic development and access to vital public services (Sadiq et al., 2023; and Taghvaee et al., 2023).

There is consensus among policymakers and economists that there are strong interrelationships between transportation and economic growth. However, the direction of causality between these variables is rather unclear in the literature. There are studies that argue that good quality institutions and the adoption of information and communication technologies (ICT) help countries to get better return on value (ROV) from their infrastructure investments. The studies show that investments in improving institutional governance and critical infrastructure (such as the transportation network) are important drivers for creating vibrant economic ecosystems and improving the quality of life of people (Owen, 1959; Rodrik et al., 2004; Ben Ali & Sassi, 2016; Zayati & Ben Ali, 2023, Zayati & Ben Ali, 2023; Ben Ali, 2023, Ben Ali, 2023; Pradhan et al., 2023a and 2023b).

The linkages between institutional quality investments in transportation infrastructure, ICT, and economic growth are rather complex. Unpacking this complexity will provide valuable insights into the types of co-development policies for ensuring LICs are able to get better ROV from these investments. To unravel the complex relationships between transportation, economic growth, institutional quality, and ICT, the study employs the panel VAR data modelling method on a sample of 79 lower income countries (LICs). The short-run and long-run dynamics derived from the empirical analysis will provide valuable insights on co-development policies pertaining to institutional governance, ICT infrastructure development, and transportation system to enable LICs to transition towards a more sustainable economic growth trajectory. The paper is organized into four remaining sections: literature and the development of hypotheses, empirical approach, results, and policy implications.

1.2 Relative Literature and Hypothesis Development

This section examines the pair-wise causal relationship between institutional quality, ICT, transportation, and economic growth. The pairwise comparison examines the existence of Granger-Causal (GC) relationships between these co-variates. In this context, we examine four major hypotheses, and they are supply-leading, demand-following, feedback, and the neutrality hypothesis.

First, we examine the link between transportation and economic growth, particularly in terms of these four hypotheses—(1) *supply-leading hypothesis* (H_{1A}), where transportation Granger-Cause (GC) economic growth (Alam et al., 2021; Zhang & Cheng, 2023; Zhu et al., 2022); (2) *demand-following hypothesis* (H_{1B}), where economic growth GC transportation (Arvin et al., 2015; Nguyen, 2023; Pradhan et al., 2021a); (3) the *feedback hypothesis* ($H_{1A,B}$), where transportation and economic growth GC one another (Pradhan, 2019; Pradhan et al., 2023b; Zhu et al., 2022); and (4) the *neutrality hypothesis*, where transportation and economic growth do not GC one another (Mohmand et al., 2021; Pradhan & Bagchi, 2013).

Second, the GC link between ICT infrastructure and economic growth, particularly in terms of these four hypotheses—(1) *supply-leading hypothesis* (H_{2A}), where ICT infrastructure GC economic growth (Pradhan et al., 2023b and 2020a); (2) *demand-following hypothesis* (H_{2B}), where economic growth GC ICT (Pradhan et al., 2021a; Nair et al., 2020; and Pradhan et al., 2022a); (3) the *feedback hypothesis* ($H_{2A,B}$), where ICT and economic growth GC one another (Pradhan et al., 2021b; and Pradhan et al., 2018); and (4) the *neutrality hypothesis*, where ICT and economic growth do not GC one another (Pradhan et al., 2023b and 2020b).

Third, the GC link between institutional quality and economic growth, particularly in terms of these four hypotheses—(1) *supply-leading hypothesis* (H_{3A}), where institutional quality GC economic growth (Pradhan et al., 2023c; Ben Ali, 2023; Ben Ali, 2023; Gurus and; Ben Ali, 2023; Ben Ali, 2023; Zayati & Ben Ali, 2023; Zayati & Ben Ali, 2023; Ben Ali et al., 2020; Ben Ali & Sorin, 2016); (2) *demand-following hypothesis* (H_{3B}), where economic growth GC institutional quality (Pradhan et al., 2022a; Nair et al., 2021; Ben Ali & Acikgoz, 2019; Swaleheen et al., 2019; Saha & Ben Ali, 2017; Ben Ali & Sassi, 2016; Ben Ali & Mdhillat, 2015); (3) the *feedback hypothesis* ($H_{3A,B}$), where institutional quality and economic growth GC one another (Pradhan et al., 2023a); and (4) the *neutrality hypothesis*, where institutional quality and economic growth do not GC one another (Pradhan et al., 2023c).

Fourth, the link between transportation GC ICT, particularly in terms of these four hypotheses—(1) *supply-leading hypothesis* (H_{4A}), where transportation GC ICT (Pradhan et al., 2023a and Adedoyin et al., 2020); (2) *demand-following hypothesis* (H_{4B}), where ICT GC transportation (Sun et al., 2022; Park et al., 2019); (3) the *feedback hypothesis* ($H_{4A,B}$), where ICT and transportation GC one another (Pradhan et al., 2023a); and (4) the *neutrality hypothesis*, where transportation and ICT do not GC one another (Pradhan et al., 2023a).

Fifth, the link between transportation and institutional quality, particularly in terms of these four hypotheses—(1) *supply-leading hypothesis* (H_{5A}), where transportation

GC institutional quality (Soh et al., 2021); (2) *demand-following hypothesis* (H_{5B}), where institutional quality GC transportation (Docherty et al., 2018); (3) the *feedback hypothesis* ($H_{5A,B}$), where institutional quality and transportation GC one another (Pradhan et al., 2023a); and (4) the *neutrality hypothesis*, where transportation and institutional quality do not GC one another (Pradhan et al., 2023a).

Sixth, the link between ICT and institutional quality, particularly in terms of these four hypotheses—(1) *supply-leading hypothesis* (H_{6A}), where ICT GC institutional quality (Pradhan et al., 2023b); (2) *demand-following hypothesis* (H_{6B}), where institutional quality GC ICT (Pradhan et al., 2023b); (3) the *feedback hypothesis* ($H_{6A,B}$), where institutional quality and ICT GC one another (Pradhan et al., 2023b); and (4) the *neutrality hypothesis*, where ICT and institutional quality do not GC one another (Pradhan et al., 2023b).

The details of these hypotheses are not highlighted here due to space restrictions. However, a summary of the literature review for these sets is presented in Table 1.1.

There has been earlier research on the foundations of these four co-variates. However, only a restricted number of studies investigate the interactions between these co-variates deploying advanced panel data techniques. In this study, we examine the short- and long-run relationship between the variables using the panel VAR model. The hypotheses tested are as follows[1]:

$H_{1A,B}$: *Transportation* ***GC*** *sustainable economic growth, and* vice versa.
$H_{2A,B}$: *ICT infrastructure* ***GC*** *sustainable economic growth, and* vice versa.
$H_{3A,B}$: *Institutional quality* ***GC*** *sustainable economic growth, and* vice versa.
$H_{4A,B}$: *Transportation* ***GC*** *ICT infrastructure, and* vice versa.
$H_{5A,B}$: *Transportation* ***GC*** *institutional quality, and* vice versa.
$H_{6A,B}$: *Institutional quality* ***GC*** *ICT infrastructure, and* vice versa.

1.3 Data, Variables, and Estimation Approach

In this study, we unpack the complex relationships between economic growth, institutional quality, ICT, and transportation infrastructure using a panel VAR model and the Granger causality (GC) technique to explore the dynamics between these four variables *simultaneously*. The hypotheses were tested for 79 LICs during the period 2005–2022. While many of these countries have lower income levels, they are rich in natural resources.[2] Additionally, several of these countries are either small islands or landlocked countries that experience unique challenges with respect to providing good transportation systems.

The data was obtained from WDI and the CPIA databases, both published by the World Bank. The variables used in the study are real per capita economic growth

[1] Detailed literature review that led to the formulation of the hypothesis can be obtained from the authors. It was not included due to space constraints.

[2] The categorization of the countries is followed on the World Bank's classification. They are not highlighted here due to space restrictions.

Table 1.1 Selected studies on the possible nexus between institutional quality, transportation, ICT infrastructure, and sustainable economic growth

Authors	Study area	Time frame	Major findings
Part A: Studies focusing on Links between Transportation and sustainable development			
Zhu et al. (2022)	China	1980–2015	F^1
Pradhan and Bagchi (2013)	India	1970–2010	F^1
Yetkiner and Beyzatlar (2020)	18 countries	1970–2017	F^1
Meersman and Nazemzadeh (2017)	Belgium	1979–2012	D^1
Mohmand et al. (2021)	Pakistan	1971–2017	F^1
Part B: Studies focusing on Links between ICT infrastructure and sustainable development			
Nair et al. (2021)	Developing countries	2005–2018	D^2, S^2, F^2
Arvin et al. (2021b)	Middle-income countries	2005–2019	F^2
Part C: Studies focusing on Links between institutional quality and sustainable development			
Nair et al. (2021)	Developing countries	2005–2018	D^3, S^3, F^3
Saidi and Mbarek (2017)	Developed countries	1990–2013	D^3
Part D: Studies focusing on Links between Transportation and ICT infrastructure			
Zergawu et al. (2020)	99 countries	1980–2015	D^4
Part E: Studies on the Link between ICT infrastructure and Institutional Quality			
Nair et al. (2021)	Developing countries	2005–2018	D^5, S^5, F^5
Karim et al. (2022)	SSA countries	2000–2021	S^5

Note 1: D stands for demand-following hypothesis; S stands for supply-leading hypothesis; F stands for feedback hypothesis; N stands for neutrality hypothesis

Note 2: D^1: Granger causality from transportation infrastructure to sustainable development only; S^1: Causality from sustainable development to transportation infrastructure only; F^1: Bidirectional Granger causality between transportation infrastructure and sustainable development; and N^1: No causality between transportation infrastructure and sustainable development

Note 3: D^2: Granger causality from institutional quality to sustainable development; S^2: Granger causality from sustainable development to institutional quality only; F^2: Bidirectional Granger causality between sustainable development and institutional quality; and N^2: No causality between institutional quality and sustainable development

Note 4: D^3: Granger causality from ICT infrastructure to sustainable development only; S^3: Granger causality from sustainable development to ICT infrastructure only; F^3: Bidirectional Granger causality between ICT infrastructure and sustainable development; and N^3: No causality between ICT infrastructure and sustainable development

Note 5: D^4: Granger causality from institutional quality to transportation infrastructure only; S^4: Granger causality from transportation infrastructure to institutional quality only; F^4: Bidirectional Granger causality between transportation infrastructure and institutional quality; and N^4: No causality between transportation infrastructure and institutional quality

Note 7: D^5: Granger causality from institutional quality to ICT infrastructure only; S^5: Granger causality from ICT infrastructure to institutional quality only; F^5: Bidirectional Granger causality between ICT infrastructure and institutional quality; and N^5: No causality between ICT infrastructure and institutional quality

Note 8: The more detailed literature relating to the linkage between institutional quality and economic growth can be explored in Pradhan et al. (2023a, b, c)

(PEG), six indicators for transportation (TRA), a composite index for institutional quality (CIQ), and a compound index for ICT (ICI). The transportation indicators are goods transported in railways (TR1), passengers carried in railways (TR2), rail lines (TR3), freight in air transport (TR4), passengers carried in air transport (TR5), and registered carrier departures worldwide in air transport (TR6).

The institutional quality (CIQ) is a composite index of 10 CPIA institutional indicators. The index was developed using PCA—principal component analysis.[3] The ICT composite index consists of six indicators.[4] The details of these indices' formulation (for both institutional quality and ICT) are not given here, but may be requested from the authors.[5] Additionally, we use CPIA indicators to arrive at the CIQ index. These include business regulatory environment rating (CBE), building human resources rating (CBH), debt policy rating (CDP), financial sector rating (CFS), economic management cluster average rating (CEM), efficiency of revenue mobilization rating (CER), macroeconomic management rating (CMM), quality of budgetary and financial management rating (CQF), quality of public administration rating (CQP), accountability, transparency, and corruption in the public sector rating (CTP).

We consider six cases based on the use of six transportation indicators (TR1–TR6). For each case, we consider 11 specifications that were based on the 11 institutional quality indicators (the 10 individual indicators and the composite index). We present a set of 6×11 relationships to capture the dynamics between the variables. We examine a set of panel regressions to capture the dynamics of the variables as shown below

$$\begin{aligned} \Delta\text{Sustainable economic growth}_{it} = \Omega_{1j} + \\ \sum_{k=1}^{q} \beta_{1ik} \Delta\text{Sustainable economic growth}_{it-k} + \sum_{k=1}^{q} \delta_{1ik} \Delta\text{Transportation}_{it-k} + \\ \sum_{k=1}^{r} \lambda_{1ik} \Delta\text{ICT Infratsructure}_{it-k} + \sum_{k=1}^{s} \mu_{1ik} \Delta\text{Institutional quality}_{it-k} + \\ \rho_{1i}\text{ECT}_{it-1} + \xi_{1it} \end{aligned} \tag{1.1}$$

$$\begin{aligned} \Delta\text{Transportation}_{it} = \Omega_{1j} + \\ \sum_{k=1}^{q} \beta_{1ik} \Delta\text{Transportation}_{it-k} + \sum_{k=1}^{q} \delta_{1ik} \Delta\text{Sustainable economic growth}_{it-k} + \\ \sum_{k=1}^{r} \lambda_{1ik} \Delta\text{ICT Infratsructure}_{it-k} + \sum_{k=1}^{s} \mu_{1ik} \Delta\text{Institutional quality}_{it-k} + \\ \rho_{1i}\text{ECT}_{it-1} + \xi_{1it} \end{aligned} \tag{1.2}$$

---

[3] The variables used to construct the CIQ are listed in the CPIA database.

[4] These include telephone landlines, mobile phones, Internet servers, Internet users, and fixed broadband automatic teller machine. The data are obtained from the WDI database.

[5] For details of this procedure, refer to Nair et al. (2021) and Pradhan et al. (2023a).

$$\begin{gathered}\Delta \text{ICT Infratsructure}_{it} = \Omega_{1j} + \\ \sum_{k=1}^{q} \beta_{1ik} \Delta \text{ICT Infratsructure}_{it-k} + \sum_{k=1}^{q} \delta_{1ik} \Delta \text{Transportation}_{it-k} + \\ \sum_{k=1}^{r} \lambda_{1ik} \Delta \text{Sustainable economic growth}_{it-k} + \sum_{k=1}^{s} \mu_{1ik} \Delta \text{Institutional quality}_{it-k} + \\ \rho_{1i} \text{ECT}_{it-1} + \xi_{1it}\end{gathered} \tag{1.3}$$

$$\begin{gathered}\Delta\Delta \text{Institutional quality}_{it} = \Omega_{1j} + \\ \sum_{k=1}^{q} \beta_{1ik} \Delta \text{Institutional quality}_{it-k} + \sum_{k=1}^{q} \delta_{1ik} \Delta \text{Transportation}_{it-k} + \\ \sum_{k=1}^{r} \lambda_{1ik} \Delta \text{ICT Infratsructure}_{it-k} + \sum_{k=1}^{s} \mu_{1ik} \Delta \text{Sustainable economic growth}_{it-k} + \\ \rho_{1i} \text{ECT}_{it-1} + \xi_{1it}\end{gathered} \tag{1.4}$$

where ECT_{-1} represents the error-correction term, indicating the long-run dynamics inherent in the set of equations, signified by 1–4. The coefficient of the first-difference variables illustrates the short-run dynamics.

1.4 Empirical Results

1.4.1 Unit Root and Cointegration Tests

The panel unit root test is first used to detect the order of integration of the data series. The test confirms that the variables are integrated into order one. Then, the panel cointegration test was undertaken, and it showed that there exists cointegration relationships between the variables. That is, there exist long-run relationships between transportation, institutional quality, ICT, and economic growth.

1.4.2 Long- and Short-Run Granger Causality Results

Using the panel vector error-correction model, we determine the possible GC relationships among transportation, ICT, institutional quality, and economic growth. Our estimation results for the six cases and 11 specifications under each case are offered in Table 3.

First, we examine the long-run GC relationship by examining the ECT_{-1} coefficients. A steady model has ECT_{-1} with negative signs, with an interval from -1 to 0. From our estimated results, we found that with Δ sustainable economic growth, the lagged ECTs were negative and significant, with a *p*-value of $p \leq 0.01$.

This suggests that economic growth converges to its long-run path in response to fluctuations in transportation, ICT, and institutional quality. The deviation from the long-run equilibrium between these variables is about 51% to 81% (see Table 1.1). This causal flow was found to be true for *all* 66 scenarios (i.e., six cases and 11 specifications under every case) that we examined. Consequently, we conclude that sustainable economic growth in LICs is significantly affected by transportation, ICT infrastructure, and institutional quality in the period under review. This suggests that long-run economic growth in LICs is dependent on the transportation system, ICT infrastructure, and institutional quality.

Next, we examine short-run causality results, which are obtained by examining the significance of the regressors of the first-difference variables. From the estimated results, we observe that the short-run inferences are non-uniform (see Table 1.2). This exhibits that the short-run dynamics vary across the six cases and eleven specifications.

Based on Table 1.2, the results regarding the various hypotheses can be discussed.

Hypothesis 1 relates to the *GC between transportation and sustainable economic growth*. For this link, 22 out of 66 specifications support the supply-leading hypothesis, indicating that sustainable economic growth drives transportation infrastructure development. This supports H_{1B} across all institutional quality indicators. The result is the same as one in the study by Kaya and Aydin (2024) for European countries using panel data during 2002–2021, Nenavath (2023) for India during 1990–2020, and Pradhan (2019) for the G-20 countries during 1961–2016. We also find that 44 out of 66 specifications support the feedback hypothesis, indicating that sustainable economic growth and transportation drive each other. This supports $H_{1A,B}$ across all institutional quality indicators. These bidirectional findings are generally consistent with the study by Nguyen (2023) for Asia using panel data during 1975–2019, Zhu et al. (2022) for China by deploying province-level panel data during 1980–2015, Mohmand et al. (2021) for Pakistan during 1971–2017, and Pradhan and Bagchi (2013) for India during 1970–2010.

For Hypothesis 2, which relates the nexus between ICT and sustainable economic growth, 22 out of 66 specifications support the demand-following hypothesis, specifying that ICT drives sustainable economic growth. This supports H_{2A} across all institutional quality indicators. The result is the same as one in the study by Pradhan et al. (2021a) for the G-20 countries during 1961–2016 and Pradhan et al. (2021b) for India using provincial data during 1991–2018. We also find that 33 out of 66 instances support the feedback hypothesis, designating that ICT and sustainable development drive each other. This supports $H_{2A,B}$ across all institutional quality indicators. These bidirectional findings are generally consistent with the study by Pradhan et al. (2022) for middle-income countries using panel data during 2005–2019, Pradhan et al. (2021b) for India using provincial data during 1991–2018, and David (2019) for Africa during 2000–2015. We also find support for the neutrality hypothesis in 11 instances, indicating that sustainable development and ICT infrastructure do not cause each other. This is the case when transportation indicators are used for rail lines only. This finding is congruent with the study by Saba et al. (2023a) for developing countries during 2000–2018, Kurniawati (2022) for Asian

Table 1.2 Results of the panel Granger causality test

Case 1: SEG, ICI, GTR, INQ											
	Specification 1 (CBE)						Specification 2 (CBH)				
	ΔSEG	ΔICI	ΔGTR	ΔCBE	ECT_{-1}		ΔSEG	ΔICI	ΔGTR	ΔCBH	ECT_{-1}
ΔSEG	–	2.62	4.67**	1.86	−0.57*	ΔSEG	–	2.56	4.69**	1.80	−0.57*
ΔICI	15.1*	–	43.1*	1.39	−0.32	ΔICI	15.2*	–	43.0*	1.41	−0.34
ΔGTR	6.14*	6.07*	–	5.05**	−0.45	ΔGTR	6.48*	6.04*	–	4.70**	−0.44
ΔCBE	7.23*	10.5*	0.40	–	−0.36	ΔCBH	6.16*	9.55*	0.51	–	−0.41
	Specification 3 (CDP)						Specification 4 (CFS)				
	ΔSEG	ΔICI	ΔGTR	ΔCDP	ECT_{-1}		ΔSEG	ΔICI	ΔGTR	ΔCFS	ECT_{-1}
ΔSEG	–	2.69	4.70**	1.02	−0.56*	ΔSEG	–	2.77	4.86**	2.56	−0.56*
ΔICI	14.6*	–	42.9*	1.21	−0.29	ΔICI	14.9*	–	43.5*	1.43	−0.30
ΔGTR	6.55*	–	6.23*	5.65**	−0.45	ΔGTR	6.66*	7.15*	–	8.68*	−0.46
ΔCDP	6.95*	9.40*	0.33	–	−0.70	ΔCES	10.8*	10.7*	1.38	–	−0.49
	Specification 5 (CEM)						Specification 6 (CER)				
	ΔSEG	ΔICI	ΔGTR	ΔCEM	ECT_{-1}		ΔSEG	ΔICI	ΔGTR	ΔCER	ECT_{-1}
ΔSEG	–	2.53	4.73**	1.15	−0.57*	ΔSEG	–	2.55	4.55**	1.14	−0.57*
ΔICI	15.0*	–	43.5*	1.39	−0.31	ΔICI	15.1*	–	43.9*	1.99	−0.31
ΔGTR	6.52*	6.14*	–	5.15**	−0.44	ΔGTR	6.37*	6.09*	–	4 94**	−0.44
ΔCEM	6.76*	11.3*	0.51	–	−0.49	ΔCER	6.06*	8.91*	0.46	–	−0.40
	Specification 7 (CMM)						Specification 8 (CQF)				
	ΔSEG	ΔICI	ΔGTR	ΔCMM	ECT_{-1}		ΔSEG	ΔICI	ΔGTR	ΔCQF	ECT_{-1}
ΔSEG	–	2.43	4.76**	1.16	−0.57*	ΔSEG	–	2.51	4.48**	1.08	−0.57*
ΔICI	15.2*	–	44.0*	1.68	−0.31	ΔICI	15.4*	–	43.3*	0.90	−0.33

(continued)

Table 1.2 (continued)

Case 1: SEG, ICI, GTR, INQ											
ΔGTR	6.51*	5.99*	–	4.69**	–0.44	ΔGTR	6.26*	6.17*	–	5.86*	–0.42
ΔCMM	6.61*	12.6*	1.41	–	–0.37	ΔCQF	8.37*	11.5*	0.75	–	–0.38
	Specification 9 (CQP)						Specification 10 (CTP)				
	ΔSEG	ΔICI	ΔGTR	ΔCQP	ECT_1		ΔSEG	ΔICI	ΔGTR	ΔCTP	ECT_1
ΔSEG	–	2.58	4.47**	1.42	–0.57*	ΔSEG	–	2.65	4.73**	2.36	–0.58*
ΔICI	15.4*	–	43.8*	1.09	–0.33	ΔICI	1.44*	–	42.6*	1.59	–0.30
ΔGTR	6.30*	5.91*	–	4.71**	–0.44	ΔGTR	6.36*	6.65*	–	4.95**	–0.43
ΔCQP	7.13*	11.3*	1.05	–	–0.30	ΔCTP	9.70*	13.9*	0.77	–	–0.29
	Specification 11 (CIQ)										
	ΔSEG	ΔICI	ΔGTR	ΔCIQ	ECT_1						
ΔSEG	–	2.55	4.64**	1.30	–0.57*						
ΔICI	15.1*	–	43.4*	1.31	–0.32						
ΔREC	6.43*	6.11*	–	4.82**	–0.45						
ΔCIQ	7.05*	9.79*	0.88	–	–1.03						
Case 2: SEG, ICI, PCR, INQ											
	Specification 1 (CBE)						Specification 2 (CBH)				
	ΔSEG	ΔICI	ΔPCR	ΔCBE	ECT_{-1}		ΔSEG	ΔICI	ΔPCR	ΔCBH	ECT_1
ΔSEG	–	2.41	11.0*	2.30	–0.59*	ΔSEG	–	2.07	10.8*	2.32	–0.59*
ΔICI	5.15**	–	23.4*	4.51**	–0.05	ΔICI	5.40**	–	23.5*	5.26**	–0.06
ΔPCR	0.77	1.03	–	1.89	–0.10	ΔPCR	0.63	0.89	–	1.87	–0.14
ΔCBE	6.63*	11.7*	6.41*	–	–0.11	ΔCBH	4.92**	10.0*	5.34*	–	–0.09
	Specification 3 (CDP)						Specification 4 (CFS)				

(continued)

Table 1.2 (continued)

Case 2: SEG, ICI, PCR, INQ											
	ΔSEG	ΔICI	ΔPCR	ΔCDP	ECT_1		ΔSEG	ΔICI	ΔPCR	ΔCFS	ECT_1
ΔSEG	–	1.99	10.1*	1.15	_0.58*	ΔSEG	–	1.92	10.6*	3.06	_0.58*
ΔICI	5.92**	–	23.7*	5.37**	_0.04	ΔICI	5.27**	–	23.9*	5.91**	_0.05
ΔPCR	0.43*	0.87	–	2.03	_0.20	ΔPCR	0.73	1.07	–	1.64	_0.13
ΔCDP	5.28*	10.4*	7.29	–	_0.41	ΔCES	7.72*	11.3*	6.42*	–	_0.04
	Specification 5 (CEM)						Specification 6 (CER)				
	ΔSEG	ΔICI	ΔPCR	ΔCEM	ECT_1		ΔSEG	ΔICI	ΔPCR	ΔCER	ECT_1
ΔSEG	–	2.10	10.5*	1.47	_0.58*	ΔSEG	–	2.17	10.9*	1.94	_0.59*
ΔICI	5.10**	–	23.9*	5.32**	_0.05	ΔICI	5.15**	–	24.4*	6.13*	_0.05
ΔPCR	0.47	0.92	–	1.77	_0.18	ΔPCR	0.58	0.95	–	1 64	_0.15
ΔCEM	5.15*	11.8*	6.41*	–	_0.22	ΔCER	4.98**	9.86*	7.17*	–	_0.04
	Specification 7 (CMM)						Specification 8 (CQF)				
	ΔSEG	ΔICI	ΔPCR	ΔCMM	ECT_1		ΔSEG	ΔICI	ΔPCR	ΔCQF	ECT_1
ΔSEG	–	2.16	10.8*	1.64	_0.58*	ΔSEG	–	2.22	10.9*	1.73	_0.59*
ΔICI	5.21**	–	24.5*	5.76**	_0.05	ΔICI	5.44*	–	23.5*	4.68**	_0.06
ΔPCR	0.50	0.95	–	1.42	_0.17	ΔPCR	0.72	0.99	–	1.15	_0.12
ΔCMM	5.53*	12.8*	7.42*	–	_0.14	ΔCQF	6.03*	11.6*	6.00*	–	_0.05
	Specification 9 (CQP)						Specification 10 (CTP)				
	ΔSEG	ΔICI	ΔPCR	ΔCQP	ECT_1		ΔSEG	ΔICI	ΔPCR	ΔCTP	ECT_1
ΔSEG	–	2.31	11.1*	2.21	_0.59*	ΔSEG	–	2.31	11.1*	2.58	_0.59*
ΔICI	5.36**	–	23.5*	4.69**	_0.05	ΔICI	5.12**	–	23.4*	7.70*	_0.04
ΔPCR	0.68	0.96	–	1.23	_0.12	ΔPCR	0.78	1.05*	–	2.29	_0.09

(continued)

Table 1.2 (continued)

Case 2: SEG, ICI, PCR, INQ											
ΔCQP	4.88**	11.6*	5.00**	–	_0.03	ΔCTP	8.11*	14.1*	5.41**	–	_0.06
	Specification 11 (CIQ)										
	ΔSEG	ΔICI	ΔPCR	ΔCIQ	ECT_1						
ΔSEG	–	2.07	10.8*	1.87	_0.59*						
ΔICI	5.46**	–	2.41*	5.89*	_0.05						
ΔPCR	0.57	0.94*	–	1.25	_0.15						
ΔCIQ	5.24*	10.5*	6.72*	–	_0.29						
Case 3: SEG, ICI, RAL, INQ											
	Specification 1 (CBE)						Specification 2 (CBH)				
	ΔSEG	ΔICI	ΔRAL	ΔCBE	ECT_{-1}		ΔSEG	ΔICI	ΔRAL	ΔCBH	ECT_1
ΔSEG	–	0.98	4.90**	1.34	_0.54*	ΔSEG	–	0.977	4.36**	1.22	_0.55*
ΔICI	0.46	–	1.25*	9.79*	_0.64	ΔICI	0.38	–	12.5*	6.21*	_0.63
ΔRAL	6.97*	6.25*	–	2.67	_0.27	ΔRAL	6.99*	6.39*	–	2.68	_0.27
ΔCBE	1.77	2.19	13.7*	–	_0.27	ΔCBH	3.08	1.70	12.8*	–	_0.49
	Specification 3 (CDP)						Specification 4 (CFS)				
	ΔSEG	ΔICI	ΔRAL	ΔCDP	ECT_1		ΔSEG	ΔICI	ΔRAL	ΔCFS	ECT_1
ΔSEG	–	1.07	4.84**	0.74	_0.52*	ΔSEG	–	1.26	4.75**	1.57	_0.51*
ΔICI	0.489	–	11.6*	4.70**	_0.79	ΔICI	0.61	–	12.4*	8.51*	_0.96
ΔRAL	6.14*	6.27*	–	3.05	_0.24	ΔRAL	4.92*	5.78*	–	2.31	_0.20
ΔCDP	6.19*	3.53	11.9*	–	_0.84	ΔCES	4.16**	4.22**	11.1*	–	_0.69
	Specification 5 (CEM)						Specification 6 (CER)				
	ΔSEG	ΔICI	ΔRAL	ΔCEM	ECT_1		ΔSEG	ΔICI	ΔRAL	ΔCER	ECT_1

(continued)

Table 1.2 (continued)

Case 3: SEG, ICI, RAL, INQ											
ΔSEG	–	0.99	4.34**	0.39	–0.53*	ΔSEG	–	0.94	5.17**	1.02	–0.54*
ΔICI	0.53	–	12.1*	6.16*	–0.83	ΔICI	0.40	–	12.7*	11.2*	–0.63
ΔRAL	6.43*	6.28*	–	3.13	–0.25	ΔRAL	6.69*	6.13*	–	2 37	–0.26
ΔCEM	4.43**	2.91	12.1*	–	–0.69	ΔCER	1.92	1.40	13.2*	–	–0.32
	Specification 7 (CMM)						Specification 8 (CQF)				
	ΔSEG	ΔICI	ΔRAL	ΔCMM	ECT_1		ΔSEG	ΔICI	ΔRAL	ΔCQF	ECT_1
ΔSEG	–	0.93	4.37**	0.63	–0.54*	ΔSEG	–	1.01	4.96**	0.67	–0.54*
ΔICI	0.44	–	12.1*	7.62*	–0.72	ΔICI	0.52	–	12.0*	5.93**	–0.76
ΔRAL	6.86	6.21*	–	3.01	–0.27	ΔRAL	6.72*	6.51*	–	2.98	–0.26
ΔCMM	2.64	2.76	11.9*	–	–0.55	ΔCQF	4.95**	1.90	13.2*	–	–0.56
	Specification 9 (CQP)						Specification 10 (CTP)				
	ΔSEG	ΔICI	ΔRAL	ΔCQP	ECT_1		ΔSEG	ΔICI	ΔRAL	ΔCTP	ECT_1
ΔSEG	–	1.08	4.68**	0.76	–0.54*	ΔSEG	–	1.09	4.66**	1.74	–0.54*
ΔICI	0.47	–	12.1*	6.67*	–0.71	ΔICI	0.30	–	12.5*	6.30*	–0.51
ΔRAL	6.92*	6.64*	–	2.45	–0.27	ΔRAL	7.42*	6.69*	–	2.67	–0.28
ΔCQP	4.51**	2.92	13.1*	–	–0.44	ΔCTP	2.62	4.64**	11.9*	–	–0.26
	Specification 11 (CIQ)										
	ΔSEG	ΔICI	ΔRAL	ΔCIQ	ECT_1						
ΔSEG	–	1.01	5.14**	0.81	–0.54*						
ΔICI	0.45	–	12.0*	6.40*	–0.75						
ΔRAL	6.57*	6.28*	–	2.74	–0.26						
ΔCIQ	5.23**	2.22	12.4*	–	–0.15						

(continued)

Table 1.2 (continued)

Case 4: SEG, ICI, RRR, INQ											
	Specification 1 (CBE)						Specification 2 (CBH)				
	ΔSEG	ΔICI	ΔRRR	ΔCBE	ECT_{-1}		ΔSEG	ΔICI	ΔRRR	ΔCBH	ECT_1
ΔSEG	–	4.69**	14.9*	0.64	_0.80*	ΔSEG	–	4.71**	15.2*	0.39	_0.81*
ΔICI	5.92*	–	3.67*	1.39*	_0.92	ΔICI	6.28*	–	3.69*	0.64	_0.95
ΔRRR	1.17	9.66	–	14.3*	_0.48	ΔRRR	1.05	10.2*	–	14.5*	_0.37
ΔCBE	0.19	5.02**	1.85	–	_0.01	ΔCBH	0.11	5.70**	1.09	–	_0.01
	Specification 3 (CDP)						Specification 4 (CFS)				
	ΔSEG	ΔICI	ΔRRR	ΔCDP	ECT_1		ΔSEG	ΔICI	ΔRRR	ΔCFS	ECT_1
ΔSEG	–	5.10**	16.1*	0.65	_0.79*	ΔSEG	–	4.64**	14.9*	0.44	_0.79*
ΔICI	6.90*	–	4.97**	3.27	_0.97	ΔICI	5.65	–	4.42**	0.54	_0.89
ΔRRR	0.97	10.1*	–	12.3*	_0.32	ΔRRR	1.11	8.73*	–	16.4*	_0.37
ΔCDP	0.04	6.36*	2.30	–	_0.04	ΔCES	0.17	6.28*	4.84**	–	_0.02
	Specification 5 (CEM)						Specification 6 (CER)				
	ΔSEG	ΔICI	ΔRRR	ΔCEM	ECT_1		ΔSEG	ΔICI	ΔRRR	ΔCER	ECT_1
ΔSEG	–	5.20**	15.4*	0.35	_0.78*	ΔSEG	–	4.89**	14.9*	0.39	_0.79*
ΔICI	7.08*	–	4.87**	2.16*	_0.99	ΔICI	6.89*	–	4.51**	0.36	_0.10
ΔRRR	1.01	9.63*	–	10.3*	_0.44	ΔRRR	1.08	9.65*	–	12 9*	_0.46
ΔCEM	0.15	6.78*	4.01	–	_0.06	ΔCER	0.04	7.39*	2.32*	–	_0.01
	Specification 7 (CMM)						Specification 8 (CQF)				
	ΔSEG	ΔICI	ΔRRR	ΔCMM	ECT_1		ΔSEG	ΔICI	ΔRRR	ΔCQF	ECT_1
ΔSEG	–	5.10**	15.2*	0.27	_0.78*	ΔSEG	–	4.78**	15.3*	0.52	_0.80*

(continued)

Table 1.2 (continued)

Case 4: SEG, ICI, RRR, INQ

ΔICI	6.98*	–	4.82**	3.38	_0.99	ΔICI	6.42*	–	4.66**	0.38	_0.96
ΔRRR	1.05	8.94*	–	8.96*	_0.48	ΔRRR	1.11	11.1*	–	19.1*	_0.39
ΔCMM	0.18	8.73*	6.95*	–	_0.06	ΔCQF	0.07	7.22**	0.89	–	_0.02
	Specification 9 (CQP)						Specification 10 (CTP)				
	ΔSEG	ΔICI	ΔRRR	ΔCQP	ECT_1		ΔSEG	ΔICI	ΔRRR	ΔCTP	ECT_1
ΔSEG	–	4.70**	15.0*	0.44	_0.81*	ΔSEG	–	4.93**	16.1*	0.49	_0.80*
ΔICI	6.32*	–	3.59	0.11	_0.96	ΔICI	5.85*	–	3.65	0.21	_0.91
ΔRRR	1.09	9.84*	–	12.3*	_0.39	ΔRRR	1.00	9.05*	–	10.5**	_0.33
ΔCQP	0.24	8.62*	2.48	–	_0.03	ΔCTP	0.08	8.04*	2.84	–	_0.02
	Specification 11 (CIQ)										
	ΔSEG	ΔICI	ΔRRR	ΔCIQ	ECT_1						
ΔSEG	–	4.89**	14.9*	0.27	_0.79*						
ΔICI	7.08	–	4.67**	0.16*	_0.11						
ΔRRR	1.06*	9.78*	–	13.8*	_0.45						
ΔCIQ	0.02	6.59*	2.08	–	_0.05						

Case 5: SEG, ICI, RRR, INQ

	Specification 1 (CBE)						Specification 2 (CBH)				
	ΔSEG	ΔICI	ΔPCA	ΔCBE	ECT_{-1}		ΔSEG	ΔICI	ΔPCA	ΔCBH	ECT_1
ΔSEG	–	4.80**	4.98**	0.28	_0.68*	ΔSEG	–	4.93**	4.99**	0.35	_0.68*
ΔICI	5.90**	–	2.84*	1.10	_0.77	ΔICI	5.15**	–	2.74	0.28	_0.80
ΔPCA	8.37*	16.0	–	6.81*	_0.62	ΔPCA	7.76*	16.3*	–	5.48**	_0.59
ΔCBE	0.30	4.96**	2.10	–	_0.04	ΔCBH	0.41	6.07*	2.59	–	_0.05

(continued)

Table 1.2 (continued)

Case 5: SEG, ICI, RRR, INQ

	Specification 3 (CDP)						Specification 4 (CFS)				
	ΔSEG	ΔICI	ΔPCA	ΔCDP	ECT_1		ΔSEG	ΔICI	ΔPCA	ΔCFS	ECT_1
ΔSEG	–	5.05**	5.32**	0.55	–0.68*	ΔSEG	–	4.77**	5.08**	0.07	–0.67*
ΔICI	5.42**	–	2.83	3.13	–0.81	ΔICI	5.07**	–	2.77	0.54	–0.78
ΔPCA	8.04*	15.2*	–	6.44*	–0.58	ΔPCA	8.08*	15.8*	–	3.43	–0.60
ΔCDP	0.20	6.16*	2.02	–	–0.01	ΔCES	0.11	5.99**	1.21	–	–0.01
	Specification 5 (CEM)						Specification 6 (CER)				
	ΔSEG	ΔICI	ΔPCA	ΔCEM	ECT_1		ΔSEG	ΔICI	ΔPCA	ΔCER	ECT_1
ΔSEG	–	5.05**	5.21**	0.41	–0.67*	ΔSEG	–	4.89**	14.9*	0.39	–0.79*
ΔICI	5.48**	–	2.81	1.10	–0.82	ΔICI	6.89*	–	4.51**	0.36	–0.10
ΔPCA	8.13	15.4*	–	4.39**	–0.60	ΔPCA	7.86*	15.7*	–	5.99**	–0.59
ΔCEM	0.09	6.95*	2.32	–	–0.01	ΔCER	0.57	7.62*	1.74*	–	–0.06
	Specification 7 (CMM)						Specification 8 (CQF)				
	ΔSEG	ΔICI	ΔPCA	ΔCMM	ECT_1		ΔSEG	ΔICI	ΔPCA	ΔCQF	ECT_1
ΔSEG	–	5.01**	5.19**	0.31	–0.66*	ΔSEG	–	4.86**	5.13**	0.32	–0.67*
ΔICI	4.47**	–	2.88**	2.05	–0.82	ΔICI	5.12**	–	2.77	0.18	–0.80
ΔPCA	8.19*	15.4*	–	5.60**	–0.61	ΔPCA	7.89*	17.1*	–	9.01*	–0.59
ΔCMM	0.20	9.22*	5.70**	–	–0.02	ΔCQF	0.24	7.31*	3.20	–	–0.04
	Specification 9 (CQP)						Specification 10 (CTP)				
	ΔSEG	ΔICI	ΔPCA	ΔCQP	ECT_1		ΔSEG	ΔICI	ΔPCA	ΔCTP	ECT_1
ΔSEG	–	4.82**	5.04**	0.14	–0.67*	ΔSEG	–	5.01**	5.41**	0.24	–0.68*
ΔICI	5.20**	–	2.76	0.10	–0.80	ΔICI	4.99**	–	2.80	0.09	–0.78

(continued)

Table 1.2 (continued)

Case 5: SEG, ICI, RRR, INQ										
ΔPCA	8.38*	14.8*	–	5.93**	–0.61	ΔPCA	7.56	14.9*	–	5.78**	–0.57
ΔCQP	0.19	8.21*	1.90	–	–0.02	ΔCTP	0.23	6.51*	0.36	–	–0.01
	Specification 11 (CIQ)										
	ΔSEG	ΔICI	ΔPCA	ΔCIQ	ECT_1						
ΔSEG	–	4.88**	4.98**	0.27	–0.67*						
ΔICI	5.25**	–	2.74	0.16*	–0.81						
ΔPCA	8.01*	15.7*	–	13.8*	–0.60						
ΔCIQ	0.35	6.61*	1.35	–	–0.11						
Case 6: SEG, ICI, RCA, INQ											
	Specification 1 (CBE)						Specification 2 (CBH)				
	ΔSEG	ΔICI	ΔRCA	ΔCBE	ECT_{-1}		ΔSEG	ΔICI	ΔRCA	ΔCBH	ECT_1
ΔSEG	–	5.07**	13.1*	0.31	–0.66*	ΔSEG	–	5.16**	12.9*	0.34	–0.67*
ΔICI	4.69**	–	3.29	0.99	–0.68	ΔICI	4.96**	–	3.31	0.29	–0.72
ΔRCA	10.4*	11.0*	–	5.05**	–0.77	ΔRCA	10.1*	11.0*	–	4.74**	–0.77
ΔCBE	0.19	4.96*	1.02	–	–0.03	ΔCBH	0.31	5.70*	3.70	–	–0.03
	Specification 3 (CDP)						Specification 4 (CFS)				
	ΔSEG	ΔICI	ΔRCA	ΔCDP	ECT_1		ΔSEG	ΔICI	ΔRCA	ΔCFS	ECT_1
ΔSEG	–	5.33**	13.6*	0.55	–0.66*	ΔSEG	–	4.88**	13.1**	0.10	–0.65*
ΔICI	4.29**	–	3.34	3.07	–0.74	ΔICI	4.87**	–	3.25	0.45	–0.71
ΔRCA	9.83*	10.9*	–	5.36**	–0.74	ΔRCA	10.2*	10.9*	–	5.74**	–0.77
ΔCDP	0.16	5.99*	2.71	–	–0.02	ΔCES	1.00	5.78**	1.43	–	–0.04
	Specification 5 (CEM)						Specification 6 (CER)				

(continued)

Table 1.2 (continued)

Case 6: SEG, ICI, RCA, INQ

	ΔSEG	ΔICI	ΔRCA	ΔCEM	ECT_1		ΔSEG	ΔICI	ΔRCA	ΔCER	ECT_1
ΔSEG	–	5.31**	13.4*	0.40	_0.65*	ΔSEG	–	5.09**	13.0*	0.13	_0.66*
ΔICI	4.31**	–	3.42	1.12	_0.75	ΔICI	4.92**	–	3.23	0.36	_0.72
ΔRCA	10.3*	10.7*	–	5.09**	_0.77	ΔRCA	9.78*	10.9*	–	4.70**	_0.76
ΔCEM	0.22	6.96*	3.85	–	_0.03	ΔCER	0.45	7.46*	1.71	–	_0.04
	Specification 7 (CMM)						Specification 8 (CQF)				
	ΔSEG	ΔICI	ΔRCA	ΔCMM	ECT_1		ΔSEG	ΔICI	ΔRCA	ΔCQF	ECT_1
ΔSEG	–	5.31**	13.3*	0.32	_0.64*	ΔSEG	–	4.86**	5.13**	0.32	_0.67*
ΔICI	4.30**	–	3.50	2.12	_0.74	ΔICI	5.12**	–	2.77	0.18	_0.80
ΔPCA	10.7*	10.7*	–	5.09**	_0.78	ΔRCA	10.2*	11.3*	–	4.95**	_0.73
ΔCMM	0.54	9.46*	8.50*	–	_0.05	ΔCQF	0.50	7.27*	3.48	–	_0.03
	Specification 9 (CQP)						Specification 10 (CTP)				
	ΔSEG	ΔICI	ΔRCA	ΔCQP	ECT_1		ΔSEG	ΔICI	ΔRCA	ΔCTP	ECT_1
ΔSEG	–	5.07**	13.2*	0.11	_0.66*	ΔSEG	–	5.27**	13.7*	0.24	_0.66*
ΔICI	4.98**	–	3.39	0.17	_0.72	ΔICI	4.82**	–	3.43	0.12	_0.70
ΔRCA	10.3*	9.98*	–	4.76**	_0.77	ΔRCA	9.32*	10.7*	–	4.72**	_0.73
ΔCQP	0.16	8.17*	2.44	–	_0.01	ΔCTP	0.28	6.29*	2.99	–	_0.03
	Specification 11 (CIQ)										
	ΔSEG	ΔICI	ΔRCA	ΔCIQ	ECT_1						
ΔSEG	–	5.10**	13.0**	0.10	_0.66*						
ΔICI	4.57**	–	3.31	0.05	_0.73						

(continued)

Table 1.2 (continued)

Case 6: SEG, ICI, RCA, INQ											
ΔRCA	10.1*	10.8*	–	5.25**	–0.77						
ΔCIQ	0.29	6.58*	2.37	–	–1.08						

Note 1: ECT_{-1} is the lagged error-correction term, and all other variables are as defined in the text
Note 2: The variable in the first column indicates the variable to which causality could possibly run
Note 3 **:** * and ** indicate the significance levels of 1% and 5%
Note 4: INQ is institutional quality, which is noted as CBE, CBH, CDP, CFS, CEM, CER, CMM, CQF, CQP, CTP, and CIQ

countries by deploying panel data during 2000–2018, and Pradhan et al. (2022) for middle-income countries using panel data during 2005–2019.

Hypothesis 3 addresses the GC between institutional quality and sustainable development. With this hypothesis, we see that 28 of 66 instances support the demand-following hypothesis, indicating that sustainable development drives institutional quality. This supports H_{3B} across all institutional quality indicators. This finding is congruent with the study by Pradhan et al., (2023a, 2023b, 2023c) for lower income countries during 2005–2022 and Law et al. (2013) for the lower income of countries during 1990–2008. We also find that 38 out of 66 instances supported the feedback hypothesis, indicating that sustainable development and institutional quality drive each other. This supports $H_{3A,B}$ across all institutional quality indicators. These bidirectional findings are generally consistent with the study by Pradhan et al., (2023a, 2023b, 2023c) for lower income countries during 2005–2022, Zayati and Ben Ali (2023) for developing countries, Pradhan et al. (2022b) for developing countries during 2000–2018, Nair et al. (2021) for developing countries during 2005–2018, and Ben Ali and Sorin (2016) for Middle East and North Africa countries.

Hypothesis 4 is concerned with the relationship between transportation and ICT infrastructure. In this instance, 22 out of 66 support the demand-following hypothesis, indicating that ICT infrastructure drives transportation. This supports H_{4B} across all institutional quality indicators. This finding is congruent with the study by Sun et al. (2022) for BRICS countries during 1990–2018 and Pradhan et al. (2021a) for the G-20 countries during 1961–2016. Additionally, 33 out of 66 instances support the feedback hypothesis, indicating that ICT and transportation drive each other. This supports $H_{4A,B}$ across all institutional quality indicators. These bidirectional findings are generally consistent with the study by Pradhan et al. (2021a) for the G-20 countries during 1961–2016 and Na et al. (2020) for 21 OECD countries during 1992–2007. We also find support for supply-leading hypothesis in 11 instances, indicating that transportation drives ICT infrastructure. This supports H_{4A} across all institutional quality indicators. This finding is congruent with the study by Pradhan et al. (2021a) for the G-20 countries during 1961–2016.

Hypothesis 5 focuses on the GC between transportation and institutional quality. In this instance, 22 out of 66 support the demand-following proposition, indicating that institutional quality drives transportation. This supports H_{5B} across all institutional quality indicators. This finding is congruent with the study by NTDPC (2013) for India, and Rietveld and Stough (2004) for the cross-national perspectives. Additionally, 44 out of 66 instances support the supply-leading proposition, indicating that transportation drives institutional quality. This supports H_{5A} across all institutional quality indicators. This finding is congruent with the study by NTDPC (2013) for India, and Rietveld and Stough (2004) for the cross-national perspectives.

Finally, we test Hypothesis 6, which relates the GC between institutional quality and ICT. In this instance, 44 out of 66 support the demand-following hypothesis, indicating that institutional quality drives ICT. This supports H_{6B} across all institutional quality indicators. This finding is congruent with the study by Pradhan et al. (2022) for middle-income countries using panel data during 2005–2019, and Evans (2022) for low-income countries during 1996–2020. We also find the support for the supply-leading hypothesis in 9 instances, indicating that ICT drives institutional quality. This uniformly supports H_{6A} across all institutional quality indicators (CBE to CIQ). This finding is congruent with the study by Pradhan et al. (2022) for middle-income countries using panel data during 2005–2019, and Evans (2022) for low-income countries during 1996–2020. Additionally, 13 out of 66 instances support the feedback proposition, showing that ICT infrastructure and institutional quality drive each other. This supports $H_{6A,B}$ across all institutional quality indicators. These bidirectional findings are generally consistent with the study by Saba et al. (2023b) for Africa during 2000–2021, and Entele (2021) for nine countries during 1995–2019.

As seen from these results, the short-run relationships are not typically uniform. They often depend on the types of transportation and/or institutional quality indicators involved. Despite these non-uniform results, overall, there were some instances of more uniform support of specific hypotheses. For instance, there is support for the feedback hypothesis for Hypothesis 1, which states that sustainable economic growth and transportation drive each other, in all 44 out of 66 instances. Similarly, there is support for the supply-leading hypothesis for Hypothesis 5, meaning that neither institutional quality drives transportation. Again, there is support for the demand-following proposition for Hypothesis 6, meaning that neither institutional quality drives ICT infrastructure. Aside from the near-uniform results for these hypotheses, there are also near-uniform results for some other hypotheses, depending on the particular case or specification used in respect of these hypotheses.

Based on the results, the overall conclusion is that there were several short-run linkages. In addition, in the long run, all these linkages pointed to the main conclusion that transportation, ICT infrastructure, and institutional quality are key drivers of sustainable economic development in middle-income countries.

1.4.3 Robustness Checks

To ensure the robustness of the empirical results, we conducted three different regression methods.[6] The Quantile-on-Quantile regression (QQR), fully modified ordinary least squares (FMOLS), dynamic OLS (DOLS), and generalized methods of moments (GMM) were undertaken. All three methods showed that transportation, ICT infrastructure, and institutional quality have a significant positive impact on sustainable economic growth. This finding is consistent with all cases and each specification.

We further employ generalized impulse response functions and the forecast error variance decomposition methods to check the power of the VECM results linking to the causal relations between transportation, ICT, institutional quality, and economic growth. The outcomes indicate the coinciding effects of shock on these variables; for instance, the mechanism identified the magnitude of shocks to economic growth

[6] The robustness test results are not presented here due to space constraint, but can be obtained from the authors.

caused by changes in transportation, ICT infrastructure, and institutional quality variables. These outcomes provide strong support for the existence of interdependence between transportation, ICT, institutional quality, and sustainable economic growth. This gives support for the findings we obtained from the VECM analysis.

1.5 Discussion, Conclusions, and Policy Implications

This paper examines the complex causal relationships among transportation, ICT infrastructure, institutional quality, and sustainable economic growth in LICs. Despite the richness of natural resources in some of the LICs, weak governance institutions and underdeveloped transportation systems hinder sustainable economic development in these economies. The study examines long-run and short-run dynamics, controlling for the state of governance institutions and ICT infrastructure.

In the short run, bidirectional and unidirectional causal relationships are observed between institutional quality, transportation, ICT, and economic growth. Different indicators related to transportation and institutional quality contribute to these interactions. In the long run, transportation systems (all six transportation systems), ICT infrastructure, and institutional quality positively influence the economic growth of LICs.

This underscores the necessity for LICs to establish a well-defined transportation plan for a smart and sustainable transportation system. Such a system, when well-integrated, has significant positive spillover effects on various economic sectors, reducing traffic congestion, enhancing business efficiency, and improving the quality of life. A key challenge faced by many LICs is access to a smart transportation system employing advanced digital governance systems. Improvements in governance systems are identified as key contributors to positive economic development in these countries.

The empirical outcomes bear crucial policy implications for LICs in transitioning to higher economic growth trajectories. The study emphasizes the importance of co-developing policies for smart transportation systems using digital technology and improvements in institutions that manage critical infrastructure in the countries. A digitally connected transportation system has the potential to reduce rent-seeking behaviour, enhance efficiency, business confidence, and attract foreign direct investments into the transportation sector and other sectors of the economy, all of which will contribute to a more dynamic and competitive economic sector in LICs.

A case study of Singapore in the 1960s serves as an illustrative example. Despite starting as a poor city-state with weak institutions, Singapore transformed itself into a globally connected and modern economy with one of the best standards of living in the world. Over six decades, the government invested in improving institutional governance and the digital architecture in the country. Today, Singapore is ranked as the top three digitally connected country in the world out of 64 countries (IMD, 2023a.) and ranked 7th in the world for government efficiency (IMD, 2023b). The island nation is one of the busiest transhipment ports in the world—it is connected to

600 ports in 123 nations and accounts for 20% of the global transhipment market (Port Technology International, 2021). These initiatives have led the gross national income (current dollars) of the country to increase from USD540 in 1965 to USD67,200 in 2022 (World Bank, 2023).

In conclusion, the paper's empirical findings underscore the strong interrelationships between institutional quality, ICT infrastructure, transportation systems, and economic growth in LICs. The empirical analysis reinforces the necessity for co-development policies focusing on developing good transportation infrastructure and governance systems utilizing digital technology. These will enhance the economic development of LICs, as shown by Singapore's success.

References

Adedoyin, F. F., Bekun, F. V., Driha, O. M., & Lorente, D. B. (2020). The effects of air transportation, energy, ICT and FDI on economic growth in the industry 4.0 era: Evidence from the United States. *Technological Forecasting and Social Change, 160*, 1–10.

Alam, K. M., Li, X., Baig, S., Ghanem, O., & Hanif, S. (2021). Causality between transportation infrastructure and economic development in Pakistan: ARDL analysis. *Research in Transportation Economics, 88*, 1–12.

Arvin, M. B., Pradhan, R. P., & Norman, N. R. (2015). Transportation intensity, urbanization, economic growth, and CO2 emissions in the G-20 countries. *Utilities Policy, 35*, 50–66.

Ben Ali, M. S. (2023). *Governance in the developing world: Transnational insights on economic development*. Springer.

Ben Ali, M. S., & Acikgoz, S. (2019). Where does economic growth in the Middle Eastern and North African countries come from? *Quarterly Review of Economics and Finance, 73*, 172–183.

Ben Ali, M. S., Fhima, F., & Nouira, R. (2020). How does corruption undermine banking stability? A threshold nonlinear framework. *Journal of Behavioral and Experimental Finance, 27*.

Ben Ali, M. S., & Mdhillat, M. (2015). Does Corruption impede international trade? New evidence from the EU and the MENA countries. *Journal of Economic Cooperation and Development, 36*(4), 107–120.

Ben Ali, M. S., & Sassi, S. (2016). The corruption-inflation nexus: evidence from developed and developing countries. *The BE Journal of Macroeconomics, 16*(1), 125–144. (Applied).

Ben Ali, M. S., Sorin M. S., K. (2016). The role of institutions in economic development. In: Ben Ali, M. S. (Eds), *Economic development in the Middle East and North Africa* (pp. 1–25). Springer/Palgrave Macmillan.

David, O. O. (2019). Nexus between telecommunication infrastructures, economic growth and development in Africa: Panel vector autoregression (P-VAR) analysis. *Telecommunications Policy, 43*, 1–17.

Docherty, I., Marseden, G., & Anable, J. (2018). The governance of smart mobility. *Transportation Research Part A, 115*, 114–125.

Entele, B. R. (2021). Impact of institutions and ICT services in avoiding resource curse: Lessons from the successful economies. *Heliyon, 7*, 1–13.

Evans, O. (2022). ICT and the provision of social services in low-income countries: The moderating role of institutional quality. *Journal of Enterprising Communities: People and Places in the Global Economy, 17*, 875–899.

Gurus, M. S., & Ben Ali, M. S. (2023). Do governance indicators predict inequality? A panel non-causality framework for the emerging countries. In: Ben Ali, M. S., & Ben Mim, S. (Eds), *Governance in the developing World: Transnational insights on economic development* (pp. 43–59). Springer.

IMD. (2023a). World Digital Competitiveness Ranking 2023. IMD, https://www.imd.org/centers/wcc/world-competitiveness-center/rankings/world-digital-competitiveness-ranking/.

IMD. (2023b). World Competitiveness Ranking 2023, IMD, https://www.imd.org/centers/wcc/world-competitiveness-center/rankings/world-competitiveness-ranking/.

Karim, S., Appiah, M., Naeem, M. A., Lucey, B. M., & Li, M. (2022). Modelling the role of institutional quality on carbon emissions in Sub-Saharan African Countries. *Renewable Energy, 198*, 213–221.

Kaya, G., & Aydin, U. (2024). The nexus between air-transport and economic growth geographically: An evidence based on heterogeneous panel data models. *Journal of Air Transport Management, 115*, 1–8.

Kurniawati, M. A. (2022). Analysis of the impact of information and communication technology on economic growth: Empirical evidence from Asian Countries. *Journal of Asian Business and Economic Studies, 29*, 2–18.

Law, S. H., Lim, T. C., & Ismail, N. W. (2013). Institutions and economic development: A granger causality analysis of panel data evidence. *Economic Systems, 37*, 610–624.

Meersman, H., & Nazemzadeh, M. (2017). The contribution of transport infrastructure to economic activity: The case of Belgium. *Case Studies on Transport Policy, 5*, 316–324.

Mohmand, Y. T., Mehmood, F., Mughal, K. S., & Aslam, F. (2021). Investigating the causal relationship between transport infrastructure, economic growth and transport emissions in Pakistan. *Research in Transportation Economics, 88*, 100972.

Na, K. Y., Kim, D. H., Park, B. G., Yoon, S. W., & Yoon, C. H. (2020). ICT and transport infrastructure development: An empirical analysis of complementarity. *Applied Economics, 52*(2), 195–211.

Nair, M., Arvin, M. B., Pradhan, R. P., & Bahmani, S. (2021). Is Higher Economic Growth Possible Through Better Institutional Quality and a Lower Carbon Footprint? Evidence from Developing Countries. *Renewable Energy, 167*, 132–145.

Nair, M., Pradhan, R. P., & Arvin, M. B. (2020). Endogenous Dynamics between R&D, ICT and Economic Growth: Empirical Evidence from the OECD Countries. *Technology in Society, 62*, 1–15.

Nenavath, S. (2023). Does Transportation Infrastructure Impact Economic Growth in India? *Journal of Facilities Management, 21*, 1–15.

Nguyen, Q. H. (2023). The causality between air transport and economic growth: Empirical evidence from regions in Asia. *Research in Transportation Business and Management, 47*, 1–15.

NTDPC. (2013). Institutions for Transport System Governance. Available at http://www.aitd.net.in/NTDPC/volume2_p1/NTDPC-Vol%2002_Part1-Ch-05.pdf.

Owen, W. (1959). Transportation and economic development. *The American Economic Review, 49*, 179–187.

Park, C., Yershov, A., & Kobzev, A. (2019). ICT infrastructure co-deployment with transport and energy infrastructure in North and Central Asia. ESCAP, UNDP.

Port Technology International. (2021). In pictures: 5 busiest transhipment hubs in the world, https://www.porttechnology.org/news/in_pictures_top_5_transhipment_hubs/.

Pradhan, P. R., Nair, S. M., Arvin, B. M., & Ben Ali, M. S. (2022b). *Intertemporal linkages between government expenditure and corruption in a digital economy in digitalization and economic development: Insights from developing countries*, Ben Ali, M. S. (Eds). Routledge, Taylor and Francis.

Pradhan, R. P., Arvin, M. B., Nair, M., Hall, J. H., & Bennett, S. E. (2020b). Sustainable economic development in India: The dynamics between financial inclusion, ICT development and economic growth. *Technological Forecasting and Social Change, 169*, 1–19.

Pradhan, R. P. (2019). Investigating the causal relationship between transportation infrastructure, financial penetration and economic growth in G-20 countries. *Research in Transportation Economics, 78*, 1–10.

Pradhan, R. P., & Bagchi, T. P. (2013). Effect of transportation infrastructure on economic growth in India: The VECM approach. *Research in Transportation Economics, 38*, 139–148.

Pradhan, R. P., Arvin, M. B., & Nair, M. (2021a). Urbanization, transportation infrastructure, ICT, and economic growth: A temporal causal analysis. *Cities, 115*, 1–19.
Pradhan, R. P., Arvin, M. B., Nair, M., & Bennett, S. E. (2020a). Sustainable economic growth in the European union: The role of ICT, venture capital and innovation. *Review of Financial Economics, 32*, 34–62.
Pradhan, R. P., Arvin, M. B., Nair, M., & Bennett, S. E. (2023c). Does foreign aid affect innovation and institutional quality? *Evaluation and Program Planning, 100*, 1–12.
Pradhan, R. P., Arvin, M. B., Nair, M., Bennett, S. E., & Bahmani, S. (2022a). Some determinants and mechanics of economic growth in middle-income countries: The role of ICT infrastructure development, taxation and other macroeconomic variables. *Singapore Economic Review, 66*, 1–37.
Pradhan, R. P., Arvin, M. B., Nair, M., Hall, J., & Bennett, S. E. (2023b). Institutional development in an information-driven economy: Can ICTs enhance economic growth for low- and lower middle-income countries? *Information Technology for Development, 28*, 468–487.
Pradhan, R. P., Arvin, M. B., Nair, M., Nair, M. S., & Bennett, S. E. (2021b). Sustainable economic development in India: The dynamics between financial inclusion, ICT development, and economic growth. *Technological Forecasting and Social Change, 169*, 1–19.
Pradhan, R. P., Arvin, M. B., Nair, M. S., & Hall, J. (2023a). Institutional quality, financial development and sustainable economic development among lower-income countries. *Natural Resources Forum- A United Nations Sustainable Development Journal, 47*, 435–483.
Pradhan, R. P., Mallik, G., Bagchi, T. P., & Sharma, M. (2018). Information communication technology penetration and stock markets-growth nexus: Evidence from cross-country panel data. *International Journal of Services and Technology Management, 24*, 307–337.
Rietveld, P., & Stough, R. (2004). Institutions, regulations and sustainable transport: A cross national perspective. *Transport Reviews, 24*, 707–719.
Saba, C. S., Djemo, C. R. T., Eita, J. H., & Ngepah, N. (2023b). Towards environmental sustainability path in Africa: The critical role of ICT, renewable energy sources, ariculturalization, industrialization and institutional quality. *Energy Reports, 10*, 1–26.
Saba, C.S., Ngepah, N., & Odhiambo, N.M. (2023a). Information and communication technology (ICT), growth and development in developing regions: evidence from a comparative analysis and a new approach. *Journal of the Knowledge Economy* (forthcoming).
Sadiq, M., Moslehpour, M., Qui, R., Hieu, V. M., Duong, K. D., & Ngo, T. Q. (2023), Sharing economy benefits and sustainable development goals: Empirical evidence from the transportation industry of Vietnam, *Journal of Innovation & Knowledge, 8*, 100290
Saha, S., & Ben Ali, M. S. (2017). Corruption and economic development: New evidence from the Middle Eastern and North African Countries. *Economic Analysis and Policy, 54*, 83–95. (Applied).
Saidi, K., & Mbarek, M. B. (2017). The impact of income, trade, urbanization, and financial development on CO_2 emissions in 19 emerging economies. *Environmental Science and Pollution Research, 24*, 12748–12757.
Soh, K. L., Wong, W. P., & Tang, C. F. (2021). The role of institutions at the nexus of logistic performance and foreign direct investment in Asia. *The Asian Journal of Shipping and Logistics, 37*, 165–173.
Stoddart, H. (2011). A Pocket guide to sustainable development governance. Stakeholder Forum.
Sun, Y., Ajaz, T., & Razzaq, A. (2022). How infrastructure development and technical efficiency change caused resources consumption in BRICS countries: Analysis based on energy, transport, ICT, and financial infrastructure indices. *Resources Policy, 79*, 1–14.
Swaleheen, M., Ben Ali, M. S., & Temimi, A. (2019). Corruption and public spending on education and Health. *Applied Economics Letters, 26*(4), 321–325.
Taghvaee, V. M., Arani, A. A., Nodehi, M., Shirazi, J. K., Agheli, L., Mohammad Neshat Ghojogh, H., Salehnia, N., Mirzaee, A., Taheri, S., Mohammadi Saber, R., Faramarzi, H., Alvandi, R., & Rahbarian, H. A. (2023). Sustainable development goals: transportation, health and public policy.

Review of Economics and Political Science, 8(2), 134–161. https://doi.org/10.1108/REPS-12-2019-0168.

World Bank. (2023). GNI per capita (current US$), Singapore, https://data.worldbank.org/indicator/NY.GNP.PCAP.CD?view=chart&locations=SG.

Yetkiner, H., & Beyzatlar, M. A. (2020). The Granger-causality between wealth and transportation: A panel data approach. *Transport Policy, 97*, 19–25.

Zayati, M., & Ben Ali, M. S. (2023). Governance and growth in developing countries: New insights from a cross-regional empirical analysis. In: M. S. Ben Ali, & S. Ben Mim (Eds), *Governance in the developing world: Transnational insights on economic development*. Springer, pp 169–201.

Zergawu, Y. Z., Walle, Y. M., & Giménez-Gómez, J. M. (2020). The joint impact of infrastructure and institutions on economic growth. *Journal of Institutional Economics, 16*(4), 481–502.

Zhang, Y., & Cheng, L. (2023). The role of transport infrastructure in economic growth: Empirical evidence in the UK. *Transport Policy, 133*, 223–233.

Zhu, F., Wu, X., & Peng, W. (2022). Road transportation and economic growth in China: Granger causality analysis based on provincial panel data. *Transportation Letters, 14*, 710–772.

Chapter 2
Does Globalization Promote Green Growth? Empirical Evidence from Organisation for Economic Co-operation and Development Countries

Muhammed Sehid Gorus

Abstract When the Sustainable Development Goals (SDGs) were adopted by the United Nations in 2015, green growth was at the center of this initiative. Although many of the SDGs' guiding principles and objectives are intimately tied to green growth, Goals 8 and 13 are directly related to this concept. Even though there have been several empirical investigations that examined the factors affecting the green growth performance of the countries, the impact of globalization has been neglected, especially at disaggregated levels. The current investigation aims to fill a gap in the empirical literature by exploring the effects of globalization on green growth by considering its different dimensions: aggregate globalization, economic globalization, trade globalization, financial globalization, social globalization, and political globalization. Besides, the environmental policy stringency index is incorporated into models as a control variable. The sample consists of 33 OECD countries covering the period from 2010 to 2020. The two-step system GMM and the Dumitrescu-Hurlin (DH) panel non-causality test are employed to reveal the aforementioned relationship. The empirical findings show that the green growth performance of the countries is positively and significantly affected by globalization, at both aggregate and disaggregate levels. The findings provide significant policy suggestions for sustainable development purposes of the countries.

Keywords Environmental policy stringency · Globalization · Green growth · OECD countries · Panel non-causality · System GMM

M. S. Gorus (✉)
Faculty of Political Sciences, Department of Economics, Ankara Yildirim Beyazit University, Ankara, Türkiye
e-mail: msgorus@aybu.edu.tr

M. S. Ben Ali and E. Lechman (eds.), *Sustainable Economic Development*,
https://doi.org/10.1007/978-981-97-3767-3_2

2.1 Introduction

Globalization is a crucial concept in today's world, affecting almost all aspects of the economy, such as international trade, the industrialization process, and the consumption behavior of economic agents. Globalization can take many forms, such as economic, social, and political globalization. Its impacts on economic growth and environmental degradation have been extensively investigated by researchers in the past decade. However, whether globalization promotes green growth or not has not been adequately explored in the empirical literature.

Actually, green growth has been at the center of the Sustainable Development Goals (SDGs) when they were adopted by the United Nations in 2015. Even though many of the SDGs' fundamental principles and targets are closely related to green growth, it has a close relationship with Goal 8 (Decent Work and Economic Growth) and Goal 13 (Climate Action). Goal 8 aims to "promote sustained, inclusive and sustainable economic growth, full and productive employment and decent work for all", while Goal 13 targets to "take urgent action to combat climate change and its impacts" (United Nations, 2023). Since green growth is an emerging concept as discussed, and considered by many advanced countries, and is vital for sustainable development, this study aims to examine the impacts of globalization on green growth in 33 OECD countries from 2010 to 2020. For this purpose, this study uses several globalization proxies developed by Gygli et al. (2019): the aggregate globalization index (GLOB), economic globalization index (EG), trade globalization index (TG), financial globalization index (FG), social globalization index (SG), and political globalization index (PG). In addition, the environmental policy stringency index (EPS) of the OECD is employed as a control variable, while the Green Growth Index (GG) of the Global Green Growth Institute (GGGI) is employed as a dependent variable in this investigation. The Green Growth Index consists of four green growth dimensions: (a) efficient and sustainable resource use, (b) natural capital protection, (c) green economic opportunities, and (d) social inclusion. This index takes values between 1.00 and 100.00. Higher values indicate better green growth performance. In addition, EPS measures the tightness of environmental regulations (environmental taxes, energy taxes, emission limits, etc.), and it takes values between 0.00 and 6.00. Higher values indicate better environmental standards. This index is very comprehensive; it considers 13 environmental policy tools that are primarily related to air pollution and climate change.

Grossman and Krueger (1991) and Bu et al. (2016) categorize the impacts of globalization on environmental quality through three main channels: the scale effect, the composition effect, and the technique effect. They argue that economic globalization negatively impacts the environment through the scale effect, while it has a positive effect through the technique effect. Besides, social globalization increases the environmental quality via the technique effect. The remaining effects of all dimensions of globalization are ambiguous including total effects (Gorus & Ben Ali, 2021). Their impacts can be significantly affected by the shares of globalization indicators within index measures. This logic can be applied to green growth measures instead of environmental quality. One can expect that globalization has a significant impact on the green growth performances of countries.

This study utilizes the subindices of economic globalization in addition to its index values, namely trade globalization and financial globalization, to capture their impacts on the green growth performances of OECD countries. Copeland (2013) claims that the level of overall production rises when trade activities increase. The environment becomes contaminated as a result of an increase in trade globalization because several steps in the industrial process require energy resource use. However, it should be noted that this argument can be valid for emerging markets or countries that ignore the environmental consequences of industrial production. Since most of the developed countries have environmental awareness, a high degree of trade globalization may increase their green growth performances. They can export more eco-friendly goods, while they can import more pollution-intensive goods. Therefore, the domestic economy may be more oriented toward environmentally friendly production. In addition, financial globalization is another subindex of economic globalization. It includes many financial indicators such as investment flows, investment agreements, investment restrictions, and so on. Ramzan et al. (2023) argue that financial globalization helps to mitigate the negative environmental outcomes of economic activities by encouraging innovation and technological progress. In addition, as mentioned above, international investment plays an important role in constructing the financial globalization index. Foreign direct investment (FDI) inflows may contribute to green growth through environmentally friendly technology transfers, improvements in energy efficiency, better environmental management, etc. (Ofori et al., 2023), in addition to their macroeconomic benefits. Taken together, it is expected that a rise in economic globalization fosters green growth in OECD countries.

Social globalization is also another factor that may promote green growth. Social globalization reflects the dissemination of ideas, information, images, and people (Gygli et al., 2019). According to Bu et al. (2016), a rise in social globalization could result in increased environmental technology spillovers that improve environmental quality as a result of greater worldwide social integration. Moreover, the environmental awareness of people increases through media and Internet access, and they demand a cleaner environment from their governments. This leads to a transition to green growth rather than gray growth (environmentally unfriendly growth) for many countries, especially high-income countries. Lastly, political globalization may have a positive impact on green growth since nations participate in more international initiatives, including environment summits and meetings, and the demand for fossil fuels steadily declines (Gorus & Karagol, 2023). For instance, the European countries introduced the European Green Deal in 2020, and they aimed to become the first carbon–neutral continent in 2050. These kinds of international efforts help to achieve green growth targets of the countries.

Methodologically, this study utilizes the two-step system Generalized Method of Moments (GMM) to estimate the parameters of the independent variables. This methodology is appropriate for conducting panel data analysis when T (time span) and N (number of countries) are small and large, respectively. In this study, a new measure, the Green Growth Index, is utilized as a dependent variable, and the dataset is only available from 2010 to 2020. Besides, the sample consists of 33 OECD countries from all around the world. Thus, this methodology fits the dataset since we have a small T and a relatively large N. Also, it is assumed in this study that the green growth performance of the countries is affected by their previous (lagged) values. Therefore, this study conducts the two-step system GMM to reveal the relationship between the above-mentioned variables. Besides, the Dumitrescu-Hurlin panel non-causality test is employed to reveal the causal nexus between variables.

The following chapter takes a snapshot of green growth, globalization, and EPS for OECD countries. Section 2.3 summarizes the empirical literature on the impacts of environmental regulations and globalization on green growth. Section 2.4 presents the dataset and method employed in this study, while Sect. 2.5 exhibits the empirical findings. The last section concludes the study.

2.2 Green Growth, Globalization, and EPS: A First Look

GGGI (2017) defined green growth as "... a development approach that seeks to deliver economic growth that is both environmentally sustainable and socially inclusive." Besides, they constructed a Green Growth Index which ranges from 1.00 to 100.00. The green growth performances of the countries can be categorized as follows: 1.00–20.00 is very low, 20.00–40.00 is low, 40.00–60.00 is moderate, 60.00–80.00 is high, and 80.00–100.00 is very high. GG scores of 33 OECD countries in 2020 are displayed in Fig. 2.1. According to the figure, it is seen that the European countries performed better than the remaining OECD countries. Among them, Austria placed the first rank with a 77.78 score in 2020, while Sweden (76.63) and Denmark (76.10) took the second and the third places, respectively. Considering 33 countries, 27 of them showed a high green growth performance, while the remaining countries exhibited a moderate performance in 2020. In detail, Israel (51.62), Iceland (51.85), South Korea (54.66), Chile (57.81), Canada (59.91), and Ireland (59.95) showed the worst green growth performance among OECD countries examined. However, it is seen that Ireland and Canada are on the edge of high performance. Thus, they need to continue their efforts on sustainable development.

The EPS was introduced by Botta and Koźluk (2014), and then revised by Kruse et al. (2022), recently. This new measure is developed to capture the progress of the countries' environmental performance, to compare their advancements, and to conduct empirical research related to its impacts on other variables such as economic, social, and environmental indicators (Kruse et al., 2022). Figure 2.2 exhibits EPS scores for 33 OECD countries in 2020. According to the data, it is seen that most of these countries have higher environmental standards. Among them, France had the highest EPS level (4.89) in 2020; in other words, it had the most stringent environmental policy across all OECD countries. Besides, South Korea, Poland, and the United Kingdom, whose index values are higher than 4.00, followed France. On the contrary, it can be concluded that Greece, Spain, Luxembourg, the Netherlands, and

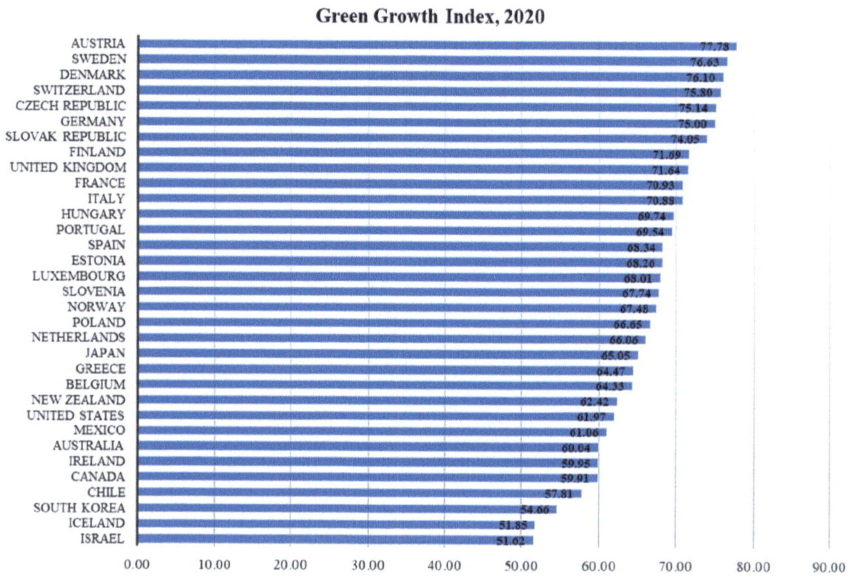

Fig. 2.1 Green Growth Index scores in 2020. *Note* The figure is constructed based on the Green Growth Index data of the GGGI

the Czech Republic had looser environmental standards. Their index values were far behind other OECD members in 2020.

The recent version of the KOF index was proposed by Gygli et al. (2019). This new globalization measure covers three main dimensions of globalization: economic, social, and political aspects. So, it is a very comprehensive index to measure the globalization level of countries in the global setting. An aggregate index for globalization for 33 OECD countries in 2020 is presented in Fig. 2.3. It is shown that these countries are highly globalized; they have an intense interaction with other countries all around the world, economically, socially, and politically. The figure demonstrates that the globalization index value is above 70.00 all over the countries. Among these countries, Switzerland (90.61), the Netherlands (90.48), and Belgium (90.09) had the highest values in 2020, while Iceland (70.28), Mexico (71.21), and New Zealand (75.36) had the lowest scores. At the disaggregated level, the Netherlands (89.57), Belgium (88.87), and Ireland's (87.98) index values were dominated by economic globalization. Considering social globalization, Luxembourg (90.18), Switzerland (90.05), and Canada (89.56) had the highest values. Besides, Italy (97.81), France (97.81), and Germany (97.60) had a high level of political globalization in 2020 compared to the remaining countries.

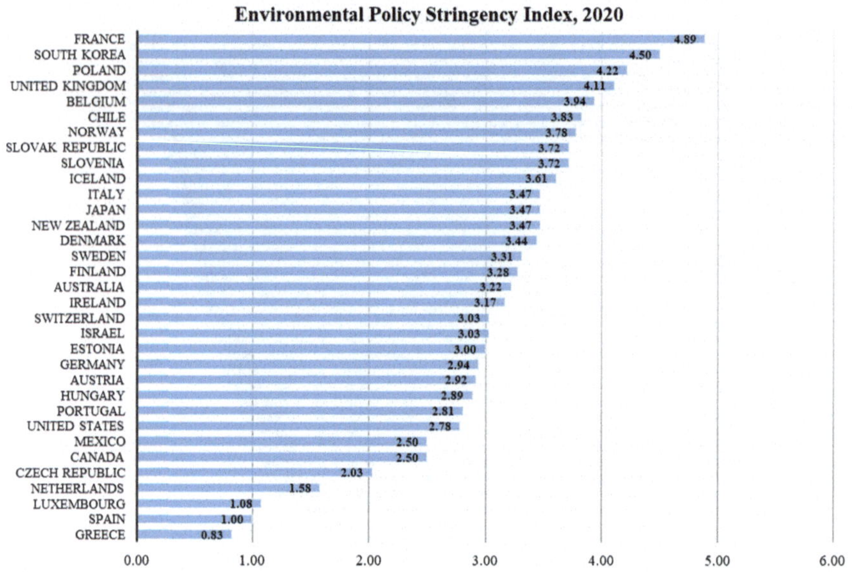

Fig. 2.2 Environmental Policy Stringency Index scores in 2020. *Note* The figure is constructed based on the EPS index data of Kruse et al. (2022)

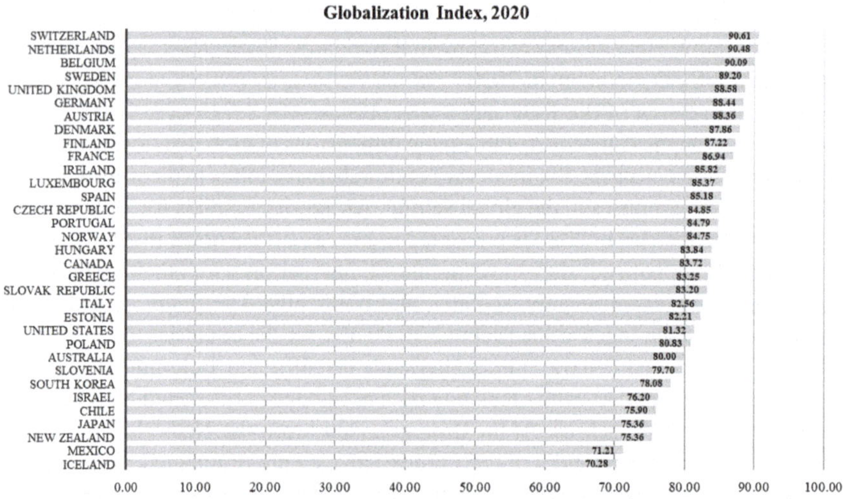

Fig. 2.3 Globalization index scores in 2020. *Note* The figure is constructed based on the KOF index data of Gygli et al. (2019)

2.3 Empirical Literature

Many empirical studies have considered the impacts of environmental policies on green growth indicators. Of the studies, Mensah et al. (2019) examined the impact of environmental taxes on green growth for 28 OECD countries covering 2000–2014. The authors employed a bunch of second-generation panel unit root and cointegration analyses. They found that high environmental taxes cause a decline in production-based CO_2 emissions for the whole panel. Besides, Jin et al. (2019) explored the impact of environmental regulations on the green total factor efficiency of industrial water resources for 30 Chinese provinces covering the period between 2000 and 2016. Their results demonstrated that tight environmental regulations cause a decline in efficiency at the national level; however, they do not have a statistically significant impact at the regional level.

Wang and Shao's (2019) empirical results confirmed that an increase in the EPS index fosters green growth in the G20 countries during the period 2001–2015. In addition, Wang et al.'s (2019) findings showed that tight environmental policies up to a certain level ($\leq$3.08 index value) have a positive impact on the green productivity growth in 27 OECD countries for 1990–2012. The empirical findings supported the validity of the Porter Hypothesis in the OECD countries examined. Besides, Cao et al. (2020) investigated the effects of environmental regulations on the green growth performance of the Chinese manufacturing industry for the period 2008–2015. The authors found that an increase in the standards of environmental regulations negatively affects green growth in the initial stage, after which its impact becomes positive, according to the system GMM estimation results.

Shuai and Fan (2020) examined the factors affecting green economic efficiency in the Chinese economy for the period from 2007 to 2018. According to the national level results, on the one hand, it was confirmed that the relationship between environmental regulations and green economic efficiency is U-shaped. On the other hand, region-based results indicated that the relationship between variables is positive monotonic, U-shaped, and again U-shaped for Eastern China, Central China, and Western China, respectively. Recently, Ben Lahouel et al. (2023) found that more stringent environmental policies make a positive contribution to inclusive green growth in 26 OECD countries during the period 1990–2012.

It is seen that environmental regulations have a significant effect on green growth for most economies. In addition to environmental policies, globalization is also another significant factor that contributes to sustainable development. Only a few empirical studies have aimed to examine this relationship. Amon these studies, Abro et al. (2023) examined the impacts of both financial globalization and trade globalization on green growth in Saudi Arabia covering 1972–2018. For this purpose, they conducted unit root and cointegration tests that consider multiple structural breaks. Subsequently, the authors employed the ARDL bounds-testing approach. Their empirical results revealed that the series are cointegrated in the long run. The findings suggested that financial globalization has a positive impact, while trade globalization has a negative effect on green growth in the long run. Besides, Chen et al. (2023) estimated the impact of financial globalization on green growth for BRICS countries during the period 1993–2019. Their results based on the CS-ARDL showed that it affected green growth positively for the period examined. Also, She and Mabrouk (2023) conducted the same methodology to determine the effect of globalization on green growth for BRICS economies, 1990–2020. The authors' findings supported the idea that globalization positively affects green growth. Their empirical findings were valid in both the short and long runs. In contrast, Ofori and Figari (2023) found that, based on the GMM results, economic globalization damages inclusive green growth when it interacts with government effectiveness and the rule of law for 23 African countries from 2000 to 2020.

The empirical studies show that there is a huge literature gap on whether a disaggregated level of globalization has a statistically significant impact on green growth, or not. Particularly, this study takes into account different dimensions of globalization as follows: aggregate globalization, economic globalization, trade globalization, financial globalization, social globalization, and political globalization. In addition, this study uses the environmental policy stringency index as a control variable that can be accepted as one of the main determinants of green growth.

2.4 Dataset and Estimation Strategy

This paper employs several globalization indices, the EPS index, and the green growth index to examine the impacts of globalization and EPS on green growth in 33 OECD countries covering 2010–2020. For this purpose, this investigation utilizes globalization indices developed by Gygli et al. (2019): the aggregate globalization index, economic globalization index, trade globalization index, financial globalization index, social globalization index, and political globalization index. Besides, EPS of the OECD is used as a control variable that is developed by Kruse et al. (2022), while GG which is gathered from GGGI is utilized as a dependent variable in the GMM estimation. All the variables are in their natural logarithms.

The Green Growth Index takes values between 1.00 and 100.00. Higher values indicate better green growth performance, while lower values denote worse green growth performance. Besides, EPS measures the strictness of environmental policies, and it takes values between 0.00 and 6.00. Higher values indicate tight ecological standards, while lower values denote lax environmental standards. EPS is highly thorough; it takes into account 13 environmental policy instruments, most of which are connected to air pollution and climate change. Market-based indicators (6 instruments), non-market-based indicators (4 instruments), and technology support (3 instruments) form the aggregate EPS index, with equal weights.

Table 2.1 exhibits the descriptive statistics of all the variables used in this study. These descriptive statistics consist of mean value, standard deviation, minimum value, and maximum value. Among these statistics, the political globalization index has the highest mean value, while EPS has the lowest mean value. Besides, it is seen that EPS has the highest standard deviation across all of the variables. It states that the variability is very high in this environmental indicator. Also, the minimum and the maximum values for each variable are reported in Table 2.1. All the variables consist of 363 observations. So, our dataset is a balanced panel.

This paper constructs 5 main models to find out the effects of globalization measures and EPS on green growth. The first model utilizes only EPS and the aggregate globalization index. The second model includes all main subindices of globalization: economic globalization, social globalization, and political globalization. The third and the fourth models consider subindices of economic globalization (trade globalization and financial globalization, respectively) separately, while the fifth model takes them into account jointly.

Model 1. $GG_{it} = \beta_{1,1}EPS_{it} + \beta_{12}GLOB_{it} + u_{1,it}$

Model 2. $GG_{it} = \beta_{2,1}EPS_{it} + \beta_{2,2}EG_{it} + \beta_{2,3}SG_{it} + \beta_{2,4}PG_{it} + u_{2,it}$

Model 3. $GG_{it} = \beta_{3,1}EPS_{it} + \beta_{3,2}TG_{it} + \beta_{3,3}SG_{it} + \beta_{3,4}PG_{it} + u_{3,it}$

Model 4. $GG_{it} = \beta_{4,1}EPS_{it} + \beta_{4,2}FG_{it} + \beta_{4,3}SG_{it} + \beta_{4,4}PG_{it} + u_{4,it}$

Model 5. $GG_{it} = \beta_{5,1}EPS_{it} + \beta_{5,2}TG_{it} + \beta_{5,3}FG_{it} + \beta_{5,4}SG_{it} + \beta_{5,5}PG_{it} + u_{5,it}$

The dynamic panel estimation method is applied in the current investigation to examine the impact of independent variables on the dependent variable. For this purpose, this study mainly employs the two-step system GMM, which was developed by Arellano and Bover (1995) and Blundell and Bond (1998), to avoid the issues arising from inserting the lagged values of the dependent variable as independent variables into the model. This methodology requires small time periods and large panel units (Roodman, 2009). After, this paper conducts the panel non-causality test suggested by Dumitrescu and Hurlin (2012) to find out the causal nexus between variables. This test is a modified and extended version of the standard Granger causality test. It calculates the standard test for each panel unit and, then, takes the arithmetic mean of these individual test statistics.

Table 2.1 Descriptive statistics

Variable	Observation	Mean	Standard deviation	Minimum value	Maximum value
GG	363	4.178	0.111	3.890	4.350
EPS	363	0.991	0.457	-0.580	1.590
GLOB	363	4.415	0.068	4.170	4.510
EG	363	4.324	0.116	3.990	4.500
TG	363	4.278	0.158	3.850	4.500
FG	363	4.364	0.116	4.050	4.570
SG	363	4.432	0.065	4.160	4.520
PG	363	4.475	0.106	4.060	4.580

2.5 Empirical Results

The estimation results derived from the two-step system GMM and the DH test are presented in Tables 2.2 and 2.3, respectively.

The empirical findings based on the GMM estimation show that tighter environmental policies promote green growth in OECD countries for the period examined. In the absence of environmental regulations, firms are able to choose any kind of production technology (labor-intensive or capital-intensive), energy mix (renewable or non-renewable energy), or industry preferences (dirty industries or clean sectors). However, most of the firms operating in developed countries have faced tight environmental regulations for a long time. For this reason, their production costs increased, and many of them exited from market or moved their operations abroad. Moving operations to other countries (especially developing countries) that have lax environmental standards helps to mitigate pollution levels in home countries because the weight of the dirty industries within the total started to decline. Besides, exiting from a dirty industry may lead the firm to enter another industry, possibly a cleaner one. Therefore, one can say that these developments promote green growth when there is a rise in environmental standards.

In addition, it is seen that globalization has a positive and statistically significant impact on green growth. Technically, a 1% increase in the aggregate globalization index causes a 0.137% rise in the green growth index in OECD countries. However, taking into account aggregate globalization may not be meaningful since this concept has many dimensions such as economic, trade, financial, social, and political globalization. Therefore, each of these dimensions should be considered separately.

At the disaggregated level, a rise in economic globalization and trade globalization causes an increase in green growth, while financial globalization does not have an impact on it. Since our sample consists of OECD countries and almost all of them are developed countries, their environmental awareness is high. Therefore, trade globalization increases their green growth performances by exporting more eco-friendly goods and importing more pollution-intensive goods. Indeed, it was anticipated that by fostering innovation and technological advancement, financial globalization would help to increase environmental quality as discussed above. This advancement could be attributed to the adoption of multinational firms' cleaner production technologies by domestic companies. However, our sample consists of OECD countries whose environmental standards and technologies are relatively better than the rest of the world. Therefore, it can be said that financial globalization or new FDI inflows do not have a significant impact on green growth. Besides, our results show that social globalization has a positive impact on green growth. A worldwide social integration raises the environmental awareness of people, thanks to media and Internet access. Similarly, a 1% increase in political globalization causes a rise in green growth by around 0.90% for each model. This finding can be attributed to the fact that the demand for clean energy rises when countries begin to take in more environmental

Table 2.2 Two-step system GMM estimation results

Variables	Model 1	Model 2	Model 3	Model 4	Model 5
GG_L1	0.853^{***} (0.014)	0.830^{***} (0.020)	0.845^{***} (0.020)	0.860^{***} (0.023)	0.851^{***} (0.022)
EPS	0.008^{***} (0.000)	0.005^{***} (0.000)	0.005^{***} (0.000)	0.004^{***} (0.000)	0.004^{***} (0.000)
GLOB	0.137^{***} (0.013)				
EG		0.031^{***} (0.011)			
TG			0.021^{***} (0.006)		0.020^{***} (0.007)
FG				0.015 (0.014)	0.012 (0.015)
SG		0.034^{**} (0.014)	0.029^{**} (0.013)	0.036^{***} (0.011)	0.018 (0.011)
PG		0.092^{***} (0.013)	0.094^{***} (0.012)	0.078^{***} (0.011)	0.088^{***} (0.012)
Arellano-Bond test for AR(1)	−3.86 [0.000]	−3.92 [0.000]	−3.93 [0.000]	−3.95 [0.000]	−3.95 [0.000]
Arellano-Bond test for AR(2)	1.03 [0.301]	1.09 [0.278]	1.07 [0.285]	1.12 [0.265]	1.08 [0.282]
Hansen Test Statistics	30.68 [0.200]	29.57 [0.241]	29.65 [0.238]	30.36 [0.211]	31.73 [0.166]

Note *** and ** denote the statistical significance levela at 1 and 5%, respectively. GG_L1 denotes the lagged value of the green growth index. The numbers in brackets show the standard errors and the numbers in square brackets denote p-values

Table 2.3 The DH panel non-causality test results

Direction	Test statistics	Direction	Test statistics
GLOB → GG	3.972***	GG → GLOB	2.878***
EG → GG	2.316**	GG → EG	3.612***
TG → GG	0.671	GG → TG	3.413***
FG → GG	1.177	GG → FG	5.521***
SG → GG	2.376***	GG → SG	2.030**
PG → GG	4.442**	GG → PG	3.674*
EPS → GG	1.085	GG → EPS	2.118**

Note ***, **, and * show the statistical significance levels at 1, 5, and 10%, respectively. The causal relationship between EPS and GG is tested by the Granger causality test (stacked test) instead of the DH panel non-causality test because of the estimation problems that arose. The corresponding test statistics are F-statistics

forums and conferences such as the Kyoto Protocol, the Paris Agreement, and the European Green Deal.

Table 2.3 exhibits that there is a two-way Granger causality between globalization and green growth for OECD countries, while there is only a one-way causality running from GG to EPS. At the disaggregated level, the empirical findings show that there is also bidirectional causality between EG and GG, SG and GG, and PG and GG. Besides, a unidirectional causal linkage is found from green growth to trade globalization and from green growth to financial globalization. Therefore, it can be said that the green growth performance of the countries has predictive power on the globalization level; its effects range from trade activities to financial transactions, from social interaction to political integration. Besides, its previous values can be used to predict future values of the environmental policy stringency index. Furthermore, the empirical results confirm that economic, social, and political globalization indices are good predictors of green growth for the countries examined. Thus, the globalization level of the countries should be monitored by authorities because it affects the sustainable growth performance of these countries.

Figure 2.4 displays an overview of causal linkages between green growth and the remaining independent variables. Our results support the idea that globalization and green growth have a close relationship.

This study produced results considering globalization which corroborate the findings of a great deal of the previous work in She and Mabrouk (2023) whose sample is BRICS countries. However, the findings of the current study do not support the previous results of Abro et al. (2023) for Saudi Arabia, Chen et al. (2023) for BRICS countries, and Ofori and Figari (2023) for African countries. A possible explanation for this might be that the current study is interested in OECD countries while the abovementioned studies consider non-OECD countries. Their economic and environmental characteristics differ across countries. The empirical results related to environmental regulations are in line with the findings of Mensah et al. (2019), Wang

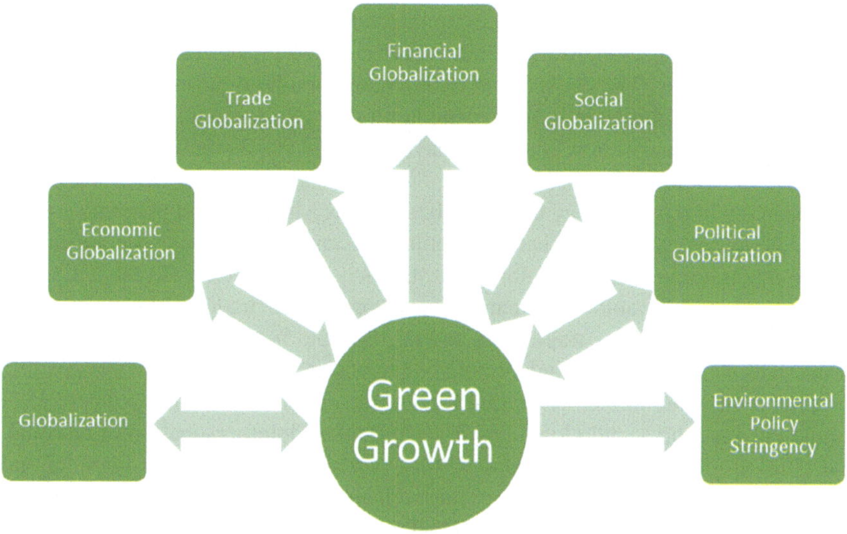

Fig. 2.4 An overview of panel causality. *Note* The figure is constructed by the author. → denotes the unidirectional causality, while ↔ represents the bidirectional causality

and Shao (2019), Wang et al. (2019), and Ben Lahouel et al. (2023). An increase in the tightness of the environmental policies helps to provide sustainable development.

2.6 Conclusion and Policy Implications

This study explored the relationship between globalization, environmental policy stringency, and green growth in 33 OECD countries for the period 2010–2020. For this purpose, the two-step system GMM and the Dumitrescu-Hurlin panel non-causality test are employed. According to the empirical findings, it was found that all dimensions of globalization (aggregate, economic, trade, social, and political), except for financial one, have a positive impact on green growth. Besides, the empirical findings showed that an increase in EPS causes a rise in the green growth index in the countries examined. Moreover, the DH panel non-causality test results exhibited that globalization and green growth have mutually affected each other for aggregate, economic, social, and political globalization. In addition, it was found that EPS has a predictive power on the green growth index.

The positive impact of trade globalization on green growth can be explained by environmental awareness. Environmental consciousness is strong in OECD nations. Thus, trade globalization improves their green growth performances by importing more commodities that cause pollution and exporting more goods that are environmentally favorable. Besides, our findings demonstrate that social globalization benefits green growth. Thanks to media and Internet access, there is a global social integration that increases people's environmental concerns. It is also found that an increase in political globalization causes a rise in the green growth performance of OECD countries. The demand for clean energy and high environmental quality increases when nations start to participate in more environmental initiatives. Lastly, the empirical evidence shows that tight environmental standards help to expel dirty industries abroad.

Taken together, we can say that globalization has a huge impact on green growth for the countries examined. Therefore, policymakers should focus on policies that help to integrate them into the global setting, including economically, socially, and politically. In detail, authorities can try to increase their trade volumes and trade partner diversity. Besides, trade taxes and tariffs could be reduced, and the number of international trade agreements should be increased. These developments are able to develop the economic globalization level of the countries. Besides, policymakers should enhance the social globalization level through interpersonal, informational, and cultural globalization. In addition, they can participate more in international environmental collaborations. All of these efforts may help to achieve the SDGs of the United Nations, especially for Goal 8 and Goal 13; countries can perform sustainable economic growth, provide productive employment, and combat climate change.

Further studies are needed to determine whether there is an impact of market-based or non-market-based environmental policies on green growth or whether their impacts differ, or not. Thus, practitioners are required to use subindices of EPS to reveal the abovementioned impacts. Besides, utilizing emerging markets as a sample is another possible future research area instead of OECD countries because the impacts of globalization and EPS on the green growth performance of developing countries might be more attractive to investigate.

Appendix

See Table 2.4.

References

Abro, A. A., Alam, N., Murshed, M., Mahmood, H., Musah, M., & Rahman, A. A. (2023). Drivers of green growth in the Kingdom of Saudi Arabia: Can financial development promote environmentally sustainable economic growth? *Environmental Science and Pollution Research, 30*(9), 23764–23780.

Arellano, M., & Bover, O. (1995). Another look at the instrumental variable estimation of error-components models. *Journal of Econometrics, 68*(1), 29–51.

Ben Lahouel, B., Taleb, L., Managi, S., & Abaoub, N. (2023). Inclusive green growth in OECD countries: What are the impacts of stringent environmental and employment regulations? *Environmental Economics and Policy Studies*, 1–29. https://doi.org/10.1007/s10018-023-00362-4

Table 2.4. List of countries

Country name		
Australia	Greece	Norway
Austria	Hungary	Poland
Belgium	Iceland	Portugal
Canada	Ireland	Slovak Republic
Chile	Israel	Slovenia
Czech Republic	Italy	South Korea
Denmark	Japan	Spain
Estonia	Luxembourg	Sweden
Finland	Mexico	Switzerland
France	Netherlands	United Kingdom
Germany	New Zealand	United States

Blundell, R., & Bond, S. (1998). Initial conditions and moment restrictions in dynamic panel data models. *Journal of Econometrics, 87*(1), 115–143.

Botta, E., & Koźluk, T. (2014). *Measuring environmental policy stringency in OECD countries: A composite index approach*. OECD Economics Department Working Papers, No. 1177. OECD Publishing. https://doi.org/10.1787/5jxrjnc45gvg-en

Bu, M., Lin, C. T., & Zhang, B. (2016). Globalization and climate change: New empirical panel data evidence. *Journal of Economic Surveys, 30*(3), 577–595.

Cao, Y., Liu, J., Yu, Y., & Wei, G. (2020). Impact of environmental regulation on green growth in China's manufacturing industry–based on the Malmquist-Luenberger index and the system GMM model. *Environmental Science and Pollution Research, 27*, 41928–41945.

Chen, R., Ramzan, M., Hafeez, M., & Ullah, S. (2023). Green innovation-green growth nexus in BRICS: Does financial globalization matter? *Journal of Innovation & Knowledge, 8*(1), Article 100286.

Copeland, B. R. (2013). Trade and the environment. *Palgrave handbook of international trade* (pp. 423–496). Palgrave Macmillan.

Dumitrescu, E. I., & Hurlin, C. (2012). Testing for Granger non-causality in heterogeneous panels. *Economic Modelling, 29*(4), 1450–1460.

GGGI. (2017). *GGGI Refreshed Strategic Plan 2015–2020*. https://gggi.org/report/gggi-refreshed-strategic-plan-2015-2020/

Gorus, M. S., & Ben Ali, M. S. (2021). Globalization and environmental pollution: Where does the MENA region stand? In: Ben Ali, M. S. (Ed.), *Economic development in the MENA region*. Perspectives on development in the Middle East and North Africa (MENA) Region. Springer.

Gorus, M. S., & Karagol, E. T. (2023). Factors affecting per capita ecological footprint in OECD countries: Evidence from machine learning techniques. *Energy & Environment, 34*(7), 2601–2618.

Grossman, G. M., & Krueger, A. B. (1991). *Environmental impacts of a North American free trade agreement*. Working paper No. 3914. National Bureau of Economic Research, Cambridge, MA.

Gygli, S., Haelg, F., Potrafke, N., & Sturm, J. E. (2019). Publisher correction to: The KOF globalisation index–revisited. *The Review of International Organizations, 14*, 575–575.

Jin, W., Zhang, H. Q., Liu, S. S., & Zhang, H. B. (2019). Technological innovation, environmental regulation, and green total factor efficiency of industrial water resources. *Journal of Cleaner Production, 211*, 61–69.

Kruse, T., Dechezleprêtre, A., Saffar, R., & Rorert, L. (2022). *Measuring environmental policy stringency in OECD countries: An update of the OECD composite EPS indicator*. OECD Economics Department Working Papers, No. 1703, OECD Publishing. https://doi.org/10.1787/90ab82e8-en

Mensah, C. N., Long, X., Dauda, L., Boamah, K. B., Salman, M., Appiah-Twum, F., & Tachie, A. K. (2019). Technological innovation and green growth in the organization for economic cooperation and development economies. *Journal of Cleaner Production, 240*, Article 118204.
Ofori, I. K., & Figari, F. (2023). Economic globalisation and inclusive green growth in Africa: Contingencies and policy-relevant thresholds of governance. *Sustainable Development, 31*(1), 452–482.
Ofori, I. K., Gbolonyo, E. Y., & Ojong, N. (2023). Foreign direct investment and inclusive green growth in Africa: Energy efficiency contingencies and thresholds. *Energy Economics, 117*, Article 106414.
Ramzan, M., Razi, U., Quddoos, M. U., & Adebayo, T. S. (2023). Do green innovation and financial globalization contribute to the ecological sustainability and energy transition in the United Kingdom? Policy insights from a bootstrap rolling window approach. *Sustainable Development, 31*(1), 393–414.
Roodman, D. (2009). How to do xtabond2: An introduction to difference and system GMM in Stata. *The Stata Journal, 9*(1), 86–136.
She, W., & Mabrouk, F. (2023). Impact of natural resources and globalization on green economic recovery: Role of FDI and green innovations in BRICS economies. *Resources Policy, 82*, Article 103479.
Shuai, S., & Fan, Z. (2020). Modeling the role of environmental regulations in regional green economy efficiency of China: Empirical evidence from super efficiency DEA-Tobit model. *Journal of Environmental Management, 261*, Article 110227.
United Nations. (2023). *The 17 goals*. Retrieved August 25, 2023, from https://sdgs.un.org/goals
Wang, X., & Shao, Q. (2019). Non-linear effects of heterogeneous environmental regulations on green growth in G20 countries: Evidence from panel threshold regression. *Science of the Total Environment, 660*, 1346–1354.
Wang, Y., Sun, X., & Guo, X. (2019). Environmental regulation and green productivity growth: Empirical evidence on the Porter Hypothesis from OECD industrial sectors. *Energy Policy, 132*, 611–619.

Chapter 3
The Causal Relationship Between Globalization and Income Inequality in the World: Toward the Achievement of the Sustainable Development Goals

Cristian Ortiz, Jonathan Zapata, and Aldo Salinas

Abstract The tenth target of the Sustainable Development Goals envisages reducing inequality within and between countries. In this regard, this research aims to assess the relationship between globalization and income inequality in different regions of the world from 1990 to 2018. We use data from official sources and advanced econometric techniques, such as CS-ARDL models, which avoid false correlations and provide robust conclusions, while heterogeneously exploring the proposed relationship in the short and long runs. The results indicate a long-run cointegration relationship between globalization and income inequality. Specifically, globalization reduces income inequality in regions with higher incomes, but has the opposite effect in regions with weaker institutions, lower incomes and more primary production structures, at least in the short run. In the long run, we find that globalization reduces inequality in all country samples. This research also reveals that globalization has a unidirectional causality from globalization to inequality in all samples. The policy implications of these results suggest the need for inclusive financial development, the incorporation of education and innovation, and the promotion of progressive fiscal policies to reduce income concentration, improve living conditions and promote sustainable economic development, especially in less developed countries.

Keywords Inequality · Globalization · Global perspective · Panel data · SDGs

JEL codes D63 · O5 · F01 · C33 · C13

C. Ortiz (✉)
Carrera de Economía and Centro de Investigaciones Sociales y Económicas, Universidad Nacional de Loja, Loja, Ecuador
e-mail: cristian.ortiz@unl.edu.ec

J. Zapata
Carrera de Economía, Universidad Nacional de Loja, Loja, Ecuador

A. Salinas
Escuela de Negocios, Universidad de Las Américas, UDLA, Quito, Ecuador

M. S. Ben Ali and E. Lechman (eds.), *Sustainable Economic Development*,
https://doi.org/10.1007/978-981-97-3767-3_3

3.1 Introduction

Income inequality is one of the global problems that affects the proper functioning of economies and social cohesion, mainly in less developed countries, where the most worrying inequality gaps have become evident. Data from the United Nations Development Programme [UNDP] (2019) revealed that, worldwide, it is estimated that 26 people own the same amount of money as the 3.8 billion poorest people in the world. Likewise, according to data from the World Income Inequality Database [WIID] (2020), the Middle East and Latin America are the most unequal regions on the planet, where, their top decile captures 56% and 54% of the average national income, while Europe, with a concentration of 35% of income, is the most egalitarian region. This situation represents great economic and social costs, since inequality undermines productivity, leading to political problems that impede the development of countries (Argadoña, 2016; Perugini and Vladisavljevic, 2020). In addition, other "costs of inequality" have been evidenced, which are mostly non-pecuniary costs. For example, the non-pecuniary costs of inequality are directly linked to higher crime rates (Reimer et al., 2021; Wink & Zanandréa, 2019), more precarious health systems (Clouston et al., 2021; Kino & Kawachi, 2020; Martins-Filho et al., 2021), wider digital divides, lower educational levels (Berger et al., 2020; Kristal & Yaish, 2020; Ravallion, 2020), less social cohesion (Álvarez, 2020) and overall lower levels of quality of life (Stiglitz, 2012).

From this perspective, the Sustainable Development Goals (SDGs) establish in their goal number ten "Reduce inequality within and between countries". For UNDP, inequality within and between countries is a continuing cause for concern, since, despite some positive signs toward reducing inequality, inequality still persists. In a more current scenario, in the wake of the COVID-19 pandemic, inequality has increased markedly and more rapidly in all countries of the world, which is attributed to weak institutions, coupled with inefficient taxation and social protection policies. In addition, the COVID-19 pandemic has led to a significant increase in unemployment worldwide, as well as a drastic reduction in workers' incomes. These developments may further exacerbate the problem of inequality and poverty. COVID-19 also represented a threat to the progress made in gender equality and women's rights over the past decades. Combating inequality and promoting equal opportunities for all must be global priorities to overcome the challenges that the pandemic has exacerbated.

A pioneering work in the study of the relationship between inequality and globalization is that proposed by Edwards (1997) who determined that the advantages brought by this process allowed for sustained growth and greater income distribution. Based on this, several studies identified the existence of a negative relationship between the variables. However, recent evidence shows that there are diverse results. For example, Cordera (2017), Vega (2017), Vera and Poy (2017) and Wang and Lu (2020) show that only some countries have benefited from economic and social globalization, while others have not, while Antonelli and Tubiana (2020), Canh et al. (2020), Le et al. (2020) and Tang et al. (2020) point out that globalization has allowed greater diffusion of goods, services, capital and knowledge, resulting in a decrease in inequality. On the other hand, Gravina and Lanzafame (2021) found that globalization generates heterogeneous effects on inequality between countries, due to regional demographic characteristics or levels of development. Precisely, the latter study motivates us to model the relationship between inequality and globalization from a heterogeneous approach. Thus, although several studies have explored the determinants of inequality in different countries (Sehid & Ben Ali, 2021; Sehid & Ben Ali, 2023), to our knowledge, few studies have paid attention to economic, social and technological globalization by studying the short- and long-term effects, while analyzing the regional characteristics of the world.

In this sense, the objective of this research is to evaluate the impact of globalization on inequality around the world using econometric techniques that address heterogeneity bias while controlling for endogeneity bias within the equations. Our main results show that there is a heterogeneous short-run and long-run effect within countries and across countries. That is, we show that, in the regions of Latin America and the Caribbean, the Middle East and North Africa, South Asia and Africa, globalization positively affects inequality, but reduces it in the long run. This fact leads us to think that globalization has a negative effect on inequality in the short term, because countries are unable to absorb the benefits of the arrival of new capital and new technologies and, rather, globalization may be generating income concentration in certain population groups. In addition, a unidirectional causal relationship from globalization to inequality is determined in all the regions studied and we find that there is a unidirectional relationship from GDP and natural resource rent to inequality. In other words, we find that in less developed regions, GDP and extractive activities accentuate income inequality in the short term.

From our perspective, this research presents three significant contributions to the literature. First, we have employed an econometric approach that models the variability present in the relationship between inequality and globalization while controlling for endogeneity bias, thus avoiding the appearance of spurious correlations and incorrect conclusions. Second, by grouping our samples by region, we have facilitated the formulation of region-specific policies, as these regions share more marked similarities compared to the traditional approach based on income levels. Third, we have highlighted the importance of the productive structure in shaping patterns of income inequality, and we have presented short-term alternatives to mitigate such inequality.

This paper is structured as follows. Section 3.2 presents a literature review based on empirical research conducted around the world. Section 3.3 describes the data and methodology applied. Section 3.4 discusses the results with empirical evidence. Finally, Sect. 3.5 presents the conclusions and policy implications of the research.

3.2 Previous Literature Review

3.2.1 *Globalization and Inequality: A Literature Overview*

Several studies have provided solid theoretical and empirical support to investigate the link between globalization and income inequality. First, some studies have found a negative relationship between globalization and inequality. For example, Vilas (1997), Hall (1998), Stiglitz (2002), Lindert and Williamson (2003) and Dussel (2004) refer to the fact that the internationalization of goods and services resulting from globalization helps to increase production and commercialization, which in the long run mitigates inequality. Furthermore, Bodemer (1998) and Appadurai (2003) state that electronic media turn the social and cultural environment around and generate more opportunities for equality in the population.

Chenery (1975), in the theory of comparative advantages, states that the development of a country is promoted through the specialization of each nation. For Williamson (1965), the endowment of resources and comparative advantages cause a favorable evolution of income inequality. Furthermore, the Economic Commission for Latin America and the Caribbean [ECLAC] (2002) recommends that, in order to remedy globalization processes, interdependence relations and levels of equity must be improved. On the other hand, Huesca and Ochoa (2016) and Afshar et al. (2019) in their studies for African and European countries verified that technology is fundamental to reducing inequality. In the same framework, Asongu et al. (2019), according to Antonelli and Tubiana (2020) and Madsen and Strulik (2020) in their study of Mexico, Spain and Italy, exposed that technology has contributed to intensive inclusive education as knowledge is the most important driver to create better living conditions.

Melitz and Redding (2015) analyze the impact of globalization in the United States and expose that globalization affects the mix of jobs available in an economy and the pace of jobs to acquire skills, a mechanism that generates equality of opportunity and income. Rotunno and Wood (2020) and Potlogea (2018) analyze globalization and the nexus with skilled cities, and determine that globalization has helped in economic activity producing changes within and outside countries, leading to better income equality. In analyzing the relationship between inequality and growth, Breunig and Majeed (2020) in their research conclude that inequality causes a negative impact on economic growth, but by taking poverty into account they were able to find that the negative effect of income inequality on economic growth is centered among countries with a high level of poverty. Tsai (1995) concluded that FDI (as a proxy for globalization) decreases income inequality in China, India, Korea, Malaysia, Singapore and Thailand. Similarly, Lee (2010) also explores the impact of globalization on income, and their results show a significant impact of globalization on income inequality, so it can be concluded that globalization is beneficial in reducing income inequality in the long run with the Kuznets curve effect.

On the other hand, diverse literature shows a positive relationship between globalization and inequality. For example, Cordera (2017), Vega (2017) and Vera and Poy

(2017) in their analysis of Mexican, Argentine and global cases agreed that, with the advent of the liberalization of capital, goods and technology, the accumulation of wealth in certain groups occurred, leading to greater inequality. In the analyses conducted by Nissanke and Thorbecke (2010) of Latin America, it was found that globalization led to an increase in inequality, as policies were implemented that were not in line with its growth rates. Recent studies such as Law et al. (2020), Gravina and Lanzafame (2021) and Bourguignon (2017) express that innovation and income inequality are subject to the level of globalization, i.e., globalization is associated with a decrease in inequality in most advanced economies, but an increase in emerging economies. Zhong et al. (2007) used panel data for U.S. metropolitan areas for the period 1980–1990 and show that globalization has increased income inequality. Wan (2007) investigates the relationship between globalization and income inequality at the provincial level and concludes that globalization promotes income inequality in China. Kai and Hamori (2009) also showed that globalization has a positive and significant impact on income inequality in Sub-Saharan African countries.

Another interesting perspective is the multidimensional perspective of growth and development, which argues that economic growth and globalization do not necessarily translate into development and less inequality and overcomes the traditional idea of the "trickle-down effect", which assumes that economic growth raises the overall welfare of society, filtering down from higher incomes to lower incomes. However, various literature (Akinci, 2018; Greenwood & Holt, 2010; Odhiambo, 2011) has shown that the contribution of income transfer from the poor to the rich is more dominant than vice versa, and the validity of the trickle-down effect is ruled out in several countries around the world. This literature shows that growth and trade openness do not necessarily mean greater levels of equality, but in some cases may accentuate patterns of inequality.

There is also literature that shows heterogeneous effects among countries with different levels of development and income. For example, Sabi (2007) revealed that globalization tends to show a positive effect in high-income economies, while the results found no significant relationship in developing, low-income and lower middle-income economies. Çelik and Basdas (2010) on the other hand showed that globalization tends to increase the income gap in developed and developing countries, while in less developed countries globalization decreases income inequality. In the same vein, Wang and Lu (2020) studied this relationship for a panel of 75 countries during the period 1970–2012, where they found that globalization affected inequality negatively in developing countries and positively in developed countries. These results were partially confirmed by Gravina and Lanzafame (2021), who found that globalization generates heterogeneous effects on inequality, due to regional demographic characteristics, levels of development and productive structure. That is, countries with more diversified productive structures better absorb the benefits of globalization, while countries with more primary productive structures tend to be affected by the arrival of new capital and technologies, which undermine the weak national industry, at least in the short term.

In this context, diverse literature has also studied the relationship between countries' productive structures and globalization. For example, Anyanwu et al. (2021)

in their analysis for countries abundant and not abundant in natural resources found that the negative impact of income inequality on economic growth increases in countries with abundant natural resources. This is due to their undiversified productive structure, which tends to concentrate income in certain groups of the population. Similarly, Lessmann and Steinkraus (2019) in their research on the geography of natural resources and the impact of natural resources on income inequality showed that countries with greater dependence on natural resource rents have higher levels of income inequality; however, better institutions can mitigate this effect (Hartwell et al., 2019). Similarly, Mohtadi and Castells-Quintana (2021) in their study on commodity price shocks and income inequality found that commodity shocks have an impact on income inequality; however, this impact depends on the type of commodity and income inequality in each country.

3.3 Data and Methodology

3.3.1 Data

In this research, we used data from 60 countries grouped in different regions of the world within the period 1990–2018. The choice of these countries was essentially due to data availability. Specifically, we divided the countries into 7 world regions such as Latin America and the Caribbean (14 countries), North America (4 countries), East Asia and the Pacific (2 countries), Europe and Central Asia (12 countries), Middle East and North Africa (9 countries), South Asia (14 countries) and Africa (5 countries). The dependent variable is income inequality, a variable that was extracted from the Standardized World Income Inequality Database [SWIID] (Solt, 2020). SWIID currently incorporates comparable Gini indices of disposable and market income inequality for 198 countries over the period 1970–2019.

On the other hand, our independent variable of interest is globalization, a variable taken from the KOF Swiss Economic Institute database (Gygli et al., 2019). In addition, as usual, some variables widely used in income equality studies were used as control variables. For example, literature shows that the size of the economy strongly predicts inequality (Potlogea, 2018; Rotunno & Wood, 2020); even an inverted U-shape is found (Kuznets, 1955), thus controlling for this effect using Gross Domestic Product (GDP). Moreover, recent literature has that the productive structure directly influences inequality (Anyanwu et al., 2021; Lessmann & Steinkraus, 2019; Mohtadi & Castells-Quintana, 2021). For them we use as a control variable natural resource rent. Finally, given that the population pattern of larger countries with larger labor force may influence inequality (Elgin and Elveren, 2021; McManus and Johnson, 2020), therefore, we use labor force as a control variable. Table A1 (Appendix) details the variables, measures and data sources.

Table A2 (Appendix) reports the descriptive statistics of the variables under analysis, where we can observe the mean, standard deviation and their minimum and maximum values. The panel is balanced in time and space with 1740 observations, 60 countries and a period of 29 years. Global income inequality presents an average of 0.567 and globalization of 9 points as the main variables that make up the relationship. Similarly, we can determine that in all the control variables, there is greater variability between countries than within each of them in the period analyzed.

Figure A1 (Appendix) shows the evolution of the income inequality variable during the period and regions studied. Here it can be highlighted that the Latin American and Caribbean region has experienced a reduction in income inequality; this was due to the fact that between 2005 and 2015 economic growth in the region was associated with an increase in the wage share. Likewise, inequality has decreased for the African, Middle East and North African regions as these regions show that between 2000 and 2017 the African economy grew at an annual rate of 4.7%; however, there are still large inequalities. Figure A2 (Appendix) shows the globalization index for all regions, where we can see that there is an increase. This behavior is mainly attributed to the fact that, over time, globalization has altered all economic, social and political structures and technology has allowed for greater interconnectedness and globalization in all regions of the world.

3.3.2 Methodology

This research aims to study the nexus between inequality and globalization in the regions of the world. To meet this objective, first, the base regression between globalization and income inequality is estimated. The relationship presented is detailed in Eq. (3.1):

$$\mathrm{INQ}_{it} = f(\mathrm{GL}_{it}) \quad (3.1)$$

where INQ_{it} corresponds to income inequality measured on a scale from 0 to 1, with 0 being perfect equality and as it approaches 1 we have perfect inequality, which is a function of GL_{it} which represents globalization that takes values from 0 to 100; therefore, a higher score indicates a higher degree of globalization, and the term ε refers to the error or stochastic disturbance. The subscript *it* indicates the value of the country i in a period t, where $i = 1, \ldots\ldots.60$ and $t = 1990, \ldots\ldots..2018$, respectively. To make the model more robust, control variables were included, as expressed in Eq. (3.2):

$$INQ_{it} = \propto_i + GL_i + Log(GDP_{it}) + NRR_{it} + Log(LF_{it}) + \varepsilon_{it} \quad (3.2)$$

The control variables used are gross domestic product (GDP), natural resource rent (NRR) and labor force (LF). In addition, some of them were transformed into logarithms for better interpretation of the model.

Next, we performed the cross-sectional dependence test proposed by Pesaran (2004) to verify possible dependence effects among the data. The test verifies the null hypothesis that there is no cross-sectional dependence (CSD); this test is defined in Eq. (3.3):

$$CD = \sqrt{\frac{2T}{N(N-1)}}\left(\sum_{i=1}^{N-1}\sum_{j=i+1}^{N}\widehat{\rho}_{ij}\right) \tag{3.3}$$

In addition, the weak cross-sectional dependence test proposed by Pesaran (2015) is used. This test measures the strength of the cross-sectional dependence of the errors; Eq. (3.4) formalizes this test as

$$CD = \left[\frac{TN(N-1)}{2}\right]^{\frac{1}{2}} \tilde{\rho}\, N \tag{3.4}$$

After verifying the cross-sectional dependence, we used the slope homogeneity test proposed by Pesaran and Yamagata (2008), which is formalized in Eqs. (3.5) and (3.6):

$$\tilde{\Delta} = \sqrt{N}\left(\frac{N^{-1}\widehat{S} - K}{\sqrt{2k}}\right), \tag{3.5}$$

$$\tilde{\Delta}_{adj} = \sqrt{\frac{N(T+1)}{T-K-1}}\left(\frac{N^{-1}\widehat{S} - K}{\sqrt{2k}}\right) \tag{3.6}$$

If the presence of cross-sectional dependence is identified, we employ the Cross-sectional Augmented IPS (CIPS) and Cross-sectional Augmented Dickey-Fuller (CADF) panel unit root tests proposed by Pesaran (2007). These second-generation tests help us to test for the existence of unit roots and, therefore, the order of integration of the variables while controlling for cross-sectional dependence. Equation (3.7) formalizes this test:

$$\Delta INQ_{it} = \alpha_i + \phi_i INQ_{it-1} + \beta_i \overline{INQ}_{t-1} + \sum_{j=0}^{k}\theta_{ij}\Delta\overline{INQ}_{it-j} + \sum_{j=0}^{k} y_{ij} INQ_{it-1} + \varepsilon_{it} \tag{3.7}$$

Then, we perform panel cointegration tests in order to verify the presence of panel cointegration. To do so, we use the second-generation panel test proposed by Westerlund (2007), which helps us to show a methodological framework to demonstrate the existence of long-run relationships while considering cross-sectional dependence in panel data. This test is formalized in Eq. (3.8):

$$\Delta INQ_{it} = \delta_i d_t + \alpha_i \left(INQ_{it-1} - \beta_i X_{it-1}\right) + \sum_{j=1}^{pi} \alpha_{ij} INQ_{it-j} \sum_{j=-qi}^{pi} \gamma_{1j} \Delta x_{it-j} + \varepsilon_{it} \tag{3.8}$$

To estimate the short- and long-run coefficients, our research is based on the CS-ARDL approach proposed by Chudik and Pesaran (2015). The CS-ARDL model has some advantages as this considers endogeneity, slope heterogeneity and cross-sectional dependence, thus providing efficient results (Chudik & Pesaran, 2015; Chudik et al., 2016). In addition, CS-ARDL is able to address both short- and long-term CSD. The CS-ARDL equation can be estimated by the following Eq. (3.9):

$$INQ_{i,t} = \alpha_1 + \beta_1 \sum_{i=0}^{k} INQ_{t-1} + \beta_2 \sum_{i=0}^{k} GL_{t-1} + \varepsilon_t \tag{3.9}$$

Finally, we analyze the directions of causality by employing the Granger non-causality panel technique proposed by Dumitrescu and Hurlin (2012). The test allows us to investigate the causal relationships between variables and to identify the direction of causality of the variables. Moreover, the causality tests of Dumitrescu and Hurlin (2012) are robust with slope heterogeneity and cross-sectional dependence. Equation (3.10) formalizes the panel causality test as follows:

$$\Delta INQ_{it} = \alpha_i + \sum_{k=1}^{K} y_i^{(k)} \Delta INQ_{it-k} + \sum_{k=1}^{K} B_i^{(k)} \Delta x_{it-k} + \varepsilon_{it} \tag{3.10}$$

3.4 Discussion of Results

Table 3.1 presents the results of the Generalized Least Squares (GLS) regression for the selected regions for the period 1990–2018. We used Eq. (3.2) for this purpose. The GLS model was employed due to the autocorrelation and heteroscedasticity problems presented by the basic model. The results indicate that in North America, East Asia and the Pacific, and Europe and South Asia, the correlation between globalization and inequality is negative. This means that these regions are benefiting from global processes. However, we find that there are no significant correlations or positive correlations in the regions of Latin America and the Caribbean; the Middle East and North Africa; South Asia and Africa. These four regions have one fact in common: they are regions whose productive structure is primary (extractive) and therefore highly dependent on natural resource rents. This evidence is in line with Law et al. (2020), Gravina and Lanzafame (2021), Bourguignon (2017) and Wang and Lu (2020) who found that globalization reduces inequality significantly in countries with higher levels of development and more diversified structures (Anyanwu et al., 2021; Lessmann & Steinkraus, 2019).

On the other hand, the control variables show the correct signs. For example, GDP has an inverted U-shape as evidenced in various literature (Canh et al., 2020; Le et al., 2020). Natural resource rent shows a positive and significant sign in several regions such as Latin America and South Asia. Likewise, the labor force does not show significant coefficients except for the Middle East and North Africa region. Overall, our results are intuitively correct; however, finding non-significant coefficients tests an interesting research question: Does globalization not influence inequality or is it rather that the effects of globalization are so heterogeneous across countries and in the short and long runs that they tend to offset each other, resulting in non-significant coefficients? To answer this interesting question, we must expand our analysis to an approach that addresses the heterogeneity in the proposed relationship.

In order to address a heterogeneity approach, some preliminary tests must first be performed. First, Table 3.2 shows the results of the cross-sectional dependence test of Pesaran (2004) and Pesaran (2015). Our results show that, indeed, the probability in the two tests performed is less than 0.01; therefore, the null hypothesis of cross-sectional independence of the variables analyzed is rejected, and we conclude that there is a high degree of cross-correlation between the variables of the model. Therefore, in order to analyze the stationarity of the variables, second-generation panel data tests will be used later on.

Subsequently, the slope homogeneity test of Pesaran and Yamagata (2008) was applied and is presented in Table 3.3. This test starts with the null hypothesis that refers to the existence of homogeneity in the slope. The results reject this hypothesis and it is concluded that there are heterogeneous parameters or slopes. This fact confirms our suspicions: there is so much heterogeneity in the coefficients of some regions that they tend to compensate each other, resulting in non-significant coefficients.

Subsequently, second-generation unit root tests were performed to test whether the series were stationary. For this purpose, the Pesaran (2007) tests were implemented. Specifically, we used the CIPS and CADF tests. The results presented in Table 3.4 reveal that both tests do not reject the null hypothesis of unit root at their levels. Therefore, all variables are stationary at the first difference, and therefore the series are integrated of order one $I(1)$.

After verifying the stationarity of the series, the Westerlund (2007) cointegration test was performed. Table 3.5 shows that the tests reject the null hypothesis of no cointegration and confirm the existence of long-term cointegration between the variables globally and for the regions studied.

In order to analyze the proposed causal relationship, and know that there is cross-sectional dependence and slope heterogeneity, we used the CS-ARDL estimator. Table 3.6 shows the results. We can observe that there is a positive effect of globalization on inequality in the short run in the regions of Latin America and the Caribbean, the Middle East and North Africa, South Asia and Africa. These results are interesting as precisely these regions have an extractive primary structure

Table 3.1 GLS model with control variables

	Global	Latin America and the Caribbean	North America	East Asia and the Pacific	Europe and Central Asia	Middle East and North Africa	South Asia	Africa
GL	0.001 (0.04)	0.000 (0.20)	−0.032** *(4.07)	−0.023*** (4.37)	−0.001* (−1.99)	−0.001 (−0.66)	0.001*** (4.87)	0.001 (0.77)
Log (GDP)	0.015*** (8.34)	0.008 (1.16)	0.013 (1.21)	0.052* (2.31)	0.011** (3.01)	0.005 (0.83)	0.051*** (10.36)	0.004 (0.85)
Log (GDP^2)	−0.016*** (−8.39)	−0.012 (−1.59)	−0.024 (1.53)	−0.052* (1.98)	−0.054** (−3.01)	−0.0312 (0.543)	−0.123* (−1.97)	−0.021 (−0.85)
NRR	0.000* (2.26)	0.000* (1.99)	0.000 (0.15)	0.002 (1.86)	0.000 (0.87)	0.000 (0.63)	0.001** (2.77)	0.000 (0.69)
Log (LF)	0.001 (0.62)	0.017* (2.10)	−0.017 (−1.38)	−0.029 (−1.18)	0.003 (0.79)	−0.019** – 3.08)	0.050*** (7.56)	−0.010 (−1.31)
Constant	0.978*** (33.69)	0.566*** (8.92)	0.447*** (3.62)	−0.728*** (−4.15)	0.803*** (15.83)	0.812*** (8.95)	0.907*** (12.69)	0.887*** (6.99)
Wooldridge (p-value)	0.000	0.028	0.001	0.05	0.000	0.000	0.000	0.000
Wald (p-value)	0.000	0.000	0.000	0.030	0.000	0.000	0.000	0.000
N	1740	406	116	58	348	261	406	145
N_g	60	14	4	2	12	9	14	5

Note Quantities in parentheses under the estimates represent standard errors. Asterisks represent the significance level of the estimators: $^{*}p < 0.05$, $^{**}p < 0.01$, and $^{***}p < 0.001$

Table 3.2 Dependence test on cross-sections

Variables	CD-test (Pesaran, 2004)		CD-test (Pesaran, 2015)	
	Statistician	*p*-value	Statistician	*p*-value
INQ	12.35***	0.000	12.353***	0.000
GL	210.17***	0.000	210.169***	0.000
Log (GDP)	212.32***	0.000	212.323***	0.000
NRR	46.38***	0.000	46.379***	0.000
Log (LF)	185.33***	0.000	185.333***	0.000

Note ***Denotes 1% significance

Table 3.3 Pesaran and Yamagata (2008) Homogeneity test

Tests	Delta	*p*-value
Δ	47.849***	0.000
Δadj	53.729***	0.000

Note ***Denotes 1% significance

and lower levels of development (Lessmann & Steinkraus, 2019). This leads us to think that globalization impacts positively these regions, due to the fact that countries fail to absorb the benefits of the arrival of new capital and new technologies (Anyanwu et al., 2021) and rather globalization may be generating income concentration in certain groups of the population. In contrast, in more developed regions such as North America, East Asia and the Pacific and Europe and Central Asia, we observe a negative effect of globalization on inequality, in the short term, is negative and significant. These findings are in line with Wang and Lu (2020), Gravina and Lanzafame (2021) and Bourguignon (2017), who found heterogeneous effects of globalization on income inequality across countries. We also find that in the long run globalization generates a negative effect on income inequality. This fact verifies our suspicion raised in the analysis in Table 3.1, where we assumed that there was short-run and long-run heterogeneity by showing non-significant coefficients. Our results show that in certain regions globalization positively affects inequality, but reduces it in the long run. These findings are of important interest as they show that public policy should focus on mitigating these short-term positive effects of globalization on inequality, especially in primary-extractive and less developed regions where short-term scenarios are crucial to generate favorable economic conditions.

Table 3.4 Results of the second-generation unit root test

Group	Test	Income inequality	GL	Log (GDP)	NRR	Log (LF)
Levels						
Global	CIPS	−11.210	−24.120	−20.229	−22.249	−12.212
	CADF	−4.412	−2.302	−4.193	−2.322	−4.149
Latin America and the Caribbean	CIPS	−13.701	−13.019	−9.131	−12.232	−2.241
	CADF	−2.723	−2.441	−4.719	−2.311	−4.439
North America	CIPS	−2.241	−7.193	−1.129	−1.120	−4.130
	CADF	−4.203	−2.209	−2.712	−2.192	−4.320
East Asia and the Pacific	CIPS	−3.202	−2.171	−1.323	−4.432	−4.130
	CADF	−4.424	−7.190	−3.224	−2.242	−2.370
Europe and Central Asia	CIPS	−1.710	−2.279	−1.930	−10.701	−7.999
	CADF	−4.441	−4.110	−4.201	−2.129	−4.200
Middle East and North Africa	CIPS	−4.412	−9.999	−11.221	−1.270	−2.320
	CADF	−3.721	−2.441	−2.102	−4.724	−3.114
South Asia	CIPS	−11.109	−11.132	−9.172	−12.343	−9.239
	CADF	−2.202	−2.291	−4.232	−2.414	−4.422
Africa	CIPS	−2.231	−2.091	−7.017	−1.224	−2.392
	CADF	−3.472	−2.712	−4.111	−2.472	−4.322
First difference						
Global	CIPS	−17.870***	−24.780***	−20.589***	−25.249***	−15.515***
	CADF	−4.478**	−5.302***	−4.793***	−5.355***	−4.149***
Latin America and the Caribbean	CIPS	−13.607***	−13.019***	−9.131***	−12.235***	−8.247**
	CADF	−5.623***	−5.441**	−4.619***	−5.371**	−4.439***
North America	CIPS	−5.541***	−6.193***	−7.129***	−7.750***	−4.730***
	CADF	−4.803***	−5.209***	−5.675***	−5.792***	−4.350***
East Asia and the Pacific	CIPS	−3.202***	−5.767***	−1.353	−4.432***	−4.730**
	CADF	−4.424***	−6.190*	−3.284***	−5.245**	−5.360***
Europe and Central Asia	CIPS	−7.670***	−8.569***	−7.930***	−10.607***	−6.999***
	CADF	−4.447***	−4.710***	−4.501***	−5.189***	−4.200***
Middle East and North Africa	CIPS	−4.418***	−9.999***	−11.287***	−7.560***	−2.380***
	CADF	−3.621***	−5.441***	−5.702***	−4.684***	−3.114***
South Asia	CIPS	−11.709***	−11.735***	−9.768***	−12.343***	−9.239***
	CADF	−5.202***	−5.291***	−4.832***	−5.414***	−4.458**
Africa	CIPS	−2.537***	−8.091***	−6.076***	−7.254***	−5.392*
	CADF	−3.462***	−5.618***	−4.771***	−5.462***	−4.355***

Note: Asterisks in parameters mean: $*p < 0.05$, $**p < 0.01$, and $***p < 0.001$

Table 3.7 shows the results of causality between the control variables and income inequality. The non-existence of causality running from inequality to GDP in the country groups and at the global level can be explained by the fact that GDP can be explained by other variables that are not taken into account; Castelló and Doménech (2002) argues that human capital inequality measures provide more robust results than income inequality measures in the estimation of standard growth and investment

Table 3.5 Results of the Westerlund (2007) cointegration test

		GL		Log (GDP)		NRR		Log (LF)	
Group	Statistician	Value	*P*-value	Value	*P*-value	Value	*P*-value	Value	*P*-value
Global	Gt	−5.052	0.000	−5.112	0.000	−4.808	0.000	−5.121	0.000
	Ga	−35.782	0.000	−37.487	0.000	−35.153	0.000	−37.859	0.000
	Pt	−43.153	0.000	−42.817	0.000	−40.842	0.000	−42.731	0.000
	Pa	−43.128	0.000	−47.676	0.000	−41.350	0.000	−44.561	0.000
Latin America and the Caribbean	Gt	−5.219	0.000	−5.248	0.000	−5.060	0.000	−5.308	0.000
	Ga	−45.436	0.000	−49.304	0.000	−46.702	0.000	−48.417	0.000
	Pt	−18.855	0.000	−19.133	0.000	−19.265	0.000	−19.788	0.000
	Pa	−52.045	0.000	−56.728	0.000	−52.049	0.000	−57.104	0.000
North America	Gt	−4.856	0.000	−4.376	0.000	−4.523	0.000	−4.976	0.000
	Ga	−27.183	0.000	−25.321	0.000	−24.248	0.000	−27.994	0.000
	Pt	−9.133	0.000	−7.382	0.000	−8.822	0.000	−9.761	0.000
	Pa	−28.403	0.000	−23.808	0.000	−25.729	0.000	−28.775	0.000
East Asia and the Pacific	Gt	−5.535	0.000	−6.066	0.000	−5.025	0.000	−5.439	0.000
	Ga	−51.084	0.000	−49.812	0.000	−48.282	0.000	−52.323	0.000
	Pt	−9.068	0.000	−9.870	0.000	−8.103	0.000	−8.249	0.000
	Pa	−58.539	0.000	−58.642	0.000	−58.041	0.000	−60.947	0.000
Europe and Central Asia	Gt	−4.775	0.000	−4.607	0.000	−4.078	0.000	−4.508	0.000
	Ga	−32.53	0.000	−32.038	0.000	−28.985	0.000	−31.221	0.000
	Pt	−19.251	0.000	−17.827	0.000	−15.194	0.000	−17.197	0.000
	Pa	−38.374	0.000	−42.127	0.000	−33.632	0.000	−38.510	0.000
Middle East and North Africa	Gt	−3.720	0.000	−3.617	0.000	−4.076	0.000	−3.996	0.000

(continued)

Table 3.5 (continued)

		GL		Log (GDP)		NRR		Log (LF)	
Group	Statistician	Value	*P*-value	Value	*P*-value	Value	*P*-value	Value	*P*-value
	Ga	−17.842	0.000	−20.007	0.000	−20.260	0.000	−20.701	0.000
	Pt	−13.099	0.000	−15.573	0.000	−19.060	0.000	−17.590	0.000
	Pa	−21.437	0.000	−29.256	0.000	−31.375	0.000	−29.770	0.000
South Asia	Gt	−5.969	0.000	−6.491	0.000	−5.805	0.000	−6.391	0.000
	Ga	−39.783	0.000	−43.554	0.000	−39.024	0.000	−45.028	0.000
	Pt	−24.566	0.000	−24.660	0.000	−22.402	0.000	−23.808	0.000
	Pa	−43.028	0.000	−51.872	0.000	−40.135	0.000	−44.699	0.000
Africa	Gt	−3.703	0.000	−3.709	0.000	−3.319	0.000	−3.406	0.000
	Ga	−19.192	0.000	−18.685	0.000	−18.556	0.000	−18.799	0.000
	Pt	−9.040	0.000	−8.693	0.000	−7.391	0.000	−8.365	0.000
	Pa	−21.04	0.000	−21.282	0.000	−19.579	0.000	−20.520	0.000

Table 3.6 Results of CS-ARDL estimator

	Global	Latin America and the Caribbean	North America	East Asia and the Pacific	Europe and Central Asia	Middle East and North Africa	South Asia	Africa
Short-run estimates								
GL	−0.123 (0.116)	0.187*** (0.019)	−0.175*** (0.031)	−0.108*** (0.033)	−0.128** (0.048)	0.108*** (0.044)	0.131** (0.038)	0.134*** (0.016)
Log. GDP	0.152*** (0.016)	0.130*** (0.021)	0.207*** (0.037)	−0.177*** (0.030)	−0.014	−0.177*** (0.040)	−0.016 (0.059)	0.152*** (0.016)
Log. GDP^2	−0.072*** (0.009)	−0.055*** (0.011)	−0.094*** (0.017)	−0.083*** (0.027)	−0.071*** (0.020)	−0.084*** (0.027)	−0.068*** (0.018)	−0.072*** (0.016)
NRR	−0.001 (0.62)	0.022* (0.010)	0.211* (0.101)	0.229* (0.111)	0.019* (0.009)	0.019** (0.008)	0.050*** (0.006)	0.011*** (0.001)
Log. LF	−0.001 (0.62)	−0.054* (0.021)	−0.021 (−1.38)	−0.031 (−1.18)	0.004 (0.79)	−0.020** (−0.07)	0.048 (1.56)	−0.002 (0.57)
Long-run Estimates								
GL	−1.148* (0.822)	−1.183*** (0.014)	−1.181*** (0.033)	−1.102*** (0.044)	−1.131*** (0.057)	−1.105*** (0.031)	−1.149*** (0.044)	−1.167*** (0.028)
Log. GDP	0.139*** (0.014)	0.113*** (0.016)	0.182*** (0.032)	0.168*** (0.029)	0.027* (0.013)	0.122*** (0.016)	0.164*** (0.052)	0.153*** (0.031)
Log. GDP^2	−0.067*** (0.010)	−0.048*** (0.010)	−0.082*** (0.016)	−0.082** (0.031)	−0.068*** (0.018)	−0.048*** (0.010)	−0.079*** (0.016)	−0.065** (0.029)
NRR	0.011 (0.32)	0.019* (0.09)	0.014 (1.11)	0.031 (1.34)	0.003 (0.43)	0.019** (0.032)	0.053*** (0.015)	0.044 (0.232)
Log. LF	−0.011 (0.62)	−0.045* (0.023)	−0.032 (−1.54)	−0.042 (−1.11)	0.023 (0.561)	−0.021** (−0.05)	0.048 (1.53)	−0.004 (0.42)
Observations	1740	406	116	58	348	261	406	145
N_groups	60	14	4	2	12	9	14	5

The dependent variable is income inequality. The asterisks in the parameters mean: $^{*}p < 0.05$, $^{**}p < 0.01$, $^{***}p < 0.001$. Standard errors in parenthesis equations, while Castelló-Climent and Doménech (2008) in his study concludes that inequality affects per capita income when investing in education taking into account their life expectancy, which depends on the human capital of their parents. Jauch and Watzka (2016) show in their results that inequality hinders growth at low to moderate levels of financial development, but has little effect on growth at high levels of financial development.

Subsequently, we can find that there is a unidirectional causal relationship running from globalization to inequality in all the regions studied. We also find a unidirectional relationship between GDP and inequality in Latin America and the Caribbean, North America, South Asia and Africa. An interesting fact is that we find a unidirectional relationship going from natural resource income to inequality in the natural resource-dependent regions: Latin America and the Caribbean, Middle East and North Africa, and South Asia and Africa. This fact confirms that regions with productive structures more specialized in the primary sector tend to have higher levels of inequality.

3.5 Conclusion and Policy Implications

The objective of this research is to analyze the nexus between globalization and income inequality in seven regions of the world within the period 1990–2018. Our research shows that globalization maintained an increasing trend during 1990–2018 in all study regions, unlike income inequality that presented a fluctuating behavior among countries, increasing at one time and decreasing at another. This behavior occurs because the productive matrix, as well as the level of development of each group of countries, is different.

Our main results show that there is a heterogeneous effect in the short and long runs within and across countries. That is, we show that, in the regions of Latin America and the Caribbean, the Middle East and North Africa, South Asia and Africa, globalization affects inequality positively, but reduces it in the long run. This fact leads us to think that globalization negatively affects inequality in the short term, because countries are unable to absorb the benefits of the arrival of new capital and new technologies, and globalization may be generating a concentration of income in certain population groups. In addition, we determine a unidirectional causal relationship from globalization to inequality in all the regions studied and find that there is a unidirectional relationship from GDP and natural resource rent to inequality. These results provide evidence that it is globalization that influences inequality and not the other way around. This indicates that public policies focused on enhancing the benefits of globalization can have a direct impact on reducing inequality, at least in the long run. In general, we find that, in less developed regions, GDP and extractive activities accentuate income inequality in the short term.

The empirical findings of this research provide important policy recommendations. For example, our results confirm that globalization increases inequality in less developed regions with a primary-extractive production structure. This fact allows

Table 3.7 Results of the Dumitrescu and Hurlin causality test

	Group	*Z*-bar	*P*-value	Result
GL ≠> INQ	Global	−1.734	0.043	Causal relationship
	Latin America and the Caribbean	−1.463	0.040	Causal relationship
	North America	1.166	0.020	Causal relationship
	East Asia and the Pacific	−1.512	0.030	Causal relationship
	Europe and Central Asia	−1.777	0.010	Causal relationship
	Middle East and North Africa	1.123	0.000	Causal relationship
	South Asia	−1.257	0.023	Causal relationship
	Africa	1.721	0.010	Causal relationship
INQ ≠> GL	Global	1.266	0.700	No causal relationship
	Latin America and the Caribbean	−1.411	0.800	No causal relationship
	North America	1.163	0.100	No causal relationship
	East Asia and the Pacific	1.111	0.100	No causal relationship
	Europe and Central Asia	1.412	0.600	No causal relationship
	Middle East and North Africa	−1.761	0.300	No causal relationship
	South Asia	1.767	0.500	No causal relationship
	Africa	−1.141	0.300	No causal relationship
Log (GDP)≠> INQ	Global	1.167	1.000	No causal relationship
	Latin America and the Caribbean	−1.273	0.030	Causal relationship
	North America	1.455	0.040	Causal relationship
	East Asia and the Pacific	−1.765	0.100	No causal relationship
	Europe and Central Asia	1.277	1.000	No causal relationship

(continued)

Table 3.7 (continued)

	Group	*Z*-bar	*P*-value	Result
	Middle East and North Africa	−1.152	0.800	No causal relationship
	South Asia	1.167	0.040	Causal relationship
	Africa	−1.611	0.000	Causal relationship
INQ ≠> Log (GDP)	Global	1.517	0.600	No causal relationship
	Latin America and the Caribbean	−1.113	0.900	No causal relationship
	North America	−1.251	0.800	No causal relationship
	East Asia and the Pacific	−1.677	0.300	No causal relationship
	Europe and Central Asia	1.136	1.000	No causal relationship
	Middle East and North Africa	−1.716	0.600	No causal relationship
	South Asia	2.373	0.100	No causal relationship
	Africa	−1.326	0.900	No causal relationship
NRR ≠> INQ	Global	2.113	0.000	Causal relationship
	Latin America and the Caribbean	1.147	0.000	Causal relationship
	North America	−1.743	0.212	No causal relationship
	East Asia and the Pacific	−1.161	0.900	No causal relationship
	Europe and Central Asia	1.363	0.020	Causal relationship
	Middle East and North Africa	1.627	0.020	Causal relationship
	South Asia	2.247	0.000	Causal relationship
	Africa	−1.311	0.0200	Causal relationship
INQ ≠> NRR	Global	5.433	1.000	No causal relationship
	Latin America and the Caribbean	6.221	1.000	No causal relationship

(continued)

Table 3.7 (continued)

	Group	*Z*-bar	*P*-value	Result
	North America	−1.136	0.400	No causal relationship
	East Asia and the Pacific	−1.676	0.600	No causal relationship
	Europe and Central Asia	1.724	0.800	No causal relationship
	Middle East and North Africa	−1.736	0.200	No causal relationship
	South Asia	5.266	0.100	No causal relationship
	Africa	1.145	0.100	No causal relationship
Log (LF) ≠> INQ	Global	5.151	0.100	No causal relationship
	Latin America and the Caribbean	1.615	0.100	No causal relationship
	North America	−1.477	0.700	No causal relationship
	East Asia and the Pacific	4.536	0.000	Causal relationship
	Europe and Central Asia	2.675	0.000	Causal relationship
	Middle East and North Africa	1.641	0.800	No causal relationship
	South Asia	4.675	0.000	Causal relationship
	Africa	−1.773	0.600	No causal relationship
INQ ≠> Log (LF)	Global	1.647	0.100	No causal relationship
	Latin America and the Caribbean	1.622	0.400	No causal relationship
	North America	−1.157	0.800	No causal relationship
	East Asia and the Pacific	1.661	0.500	No causal relationship
	Europe and Central Asia	2.736	0.100	No causal relationship
	Middle East and North Africa	−1.172	0.300	No causal relationship

(continued)

Table 3.7 (continued)

	Group	Z-bar	P-value	Result
	South Asia	1.121	0.400	No causal relationship
	Africa	1.3716	0.900	No causal relationship

us to recommend that policy makers should focus on short-term inclusive policies that efficiently redistribute income. Moreover, diversifying economies toward more complex goods is an appropriate strategy to deconcentrate income in a few population groups. In other words, globalization should focus on generating productive linkages and the production of more complex goods. Also, distributive and tax policies should be implemented to help mitigate the short-term impact on less developed regions. These actions will ensure compliance with the tenth Sustainable Development Goal.

Acknowledgements The authors acknowledge and thank the wise comments of two anonymous referees and the financial support of the Universidad Nacional de Loja research project: 38-DI-FJSA-2023.

Appendix

See Tables A1, A2, Figs. A1 and A2.

Table A1 Description of variables

Variable	Symbol	Measure	Description	Source of data
Dependent				
Income inequality	INQ	Index	The Gini index measures the extent to which the distribution of income among individuals or households within an economy departs from a perfectly equal distribution. The index takes values between 0 and 1. Thus, a Gini index of 0 represents perfect equity, while an index of 1 represents perfect inequity	WIID (2020)
Independent				
Globalization	GL	Index	A phenomenon based on the continuous increase in the interconnection between different nations in the world at the economic, political, social and technological levels. The index takes values between 0 and 100. Thus, a globalization index of 100 means the highest value of globalization	Gygli et al., 2019
Controls				
Gross domestic product	GDP	$USD	GDP at constant prices is the gross value added by all resident producers in the economy plus any product contributions and minus any subsidies not included in the value of the products. Log for analysis	World Bank (2020)

(continued)

Table A1 (continued)

Variable	Symbol	Measure	Description	Source of data
Natural resource rentals	NRR	Percentage of GDP	Total natural resource rent is the sum of oil rent, natural gas rent, coal rent (hard and soft), mineral rent and forestry rent	World Bank (2020)
Labor force	LF	Number of persons	The workforce is the skills and knowledge that you have dedicated to your work, in a given time. Log for analysis	World Bank (2020)

Table A2 Descriptive statistics

Variable		Media	Standard deviation	Minimum	Maximum	Remarks
INQ	Global	0.567	0.089	0.299	0.781	$N = 1740$
	Enter		0.085	0.391	0.753	$n = 60$
	Inside		0.027	0.425	0.668	$T = 29$
GL	Global	60.339	16.050	22.356	89.720	$N = 1740$
	Enter		14.753	31.398	86.018	$n = 60$
	Inside		6.591	36.017	77.794	$T = 29$
Log. GDP	Global	25.355	2.170	19.590	30.517	$N = 1740$
	Enter		2.160	20.469	30.205	$n = 60$
	Inside		0.349	23.980	26.555	$T = 29$
NRR	Global	5.526	8.179	0.000	55.523	$N = 1740$
	Enter		7.469	0.001	37.087	$n = 60$
	Inside		3.464	−19.081	32.899	$T = 29$
Log. FL	Global	15.855	1.733	11.450	20.484	$N = 1740$
	Enter		1.738	11.891	20.422	$n = 60$
	Inside		0.181	15.296	16.424	$T = 29$

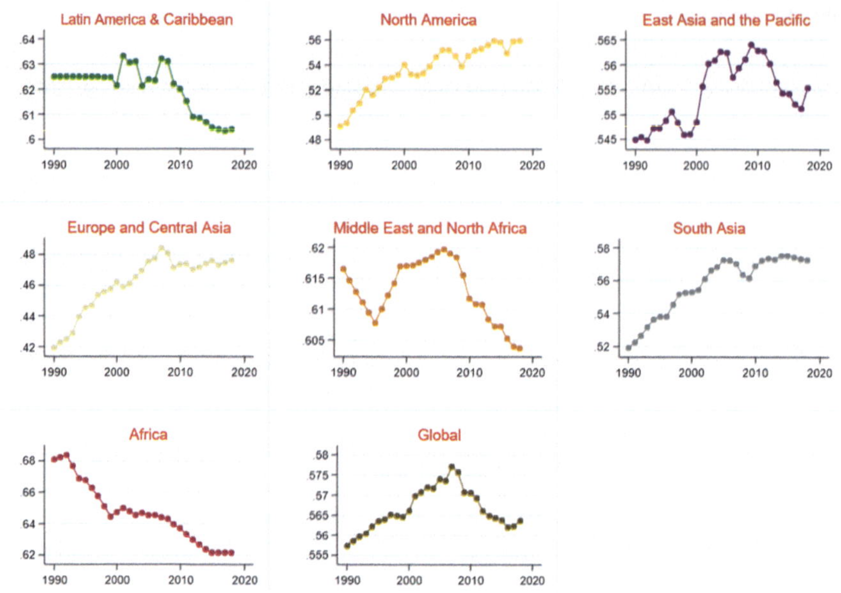

Fig. A1 Evolution of income inequality in the regions of the world studied

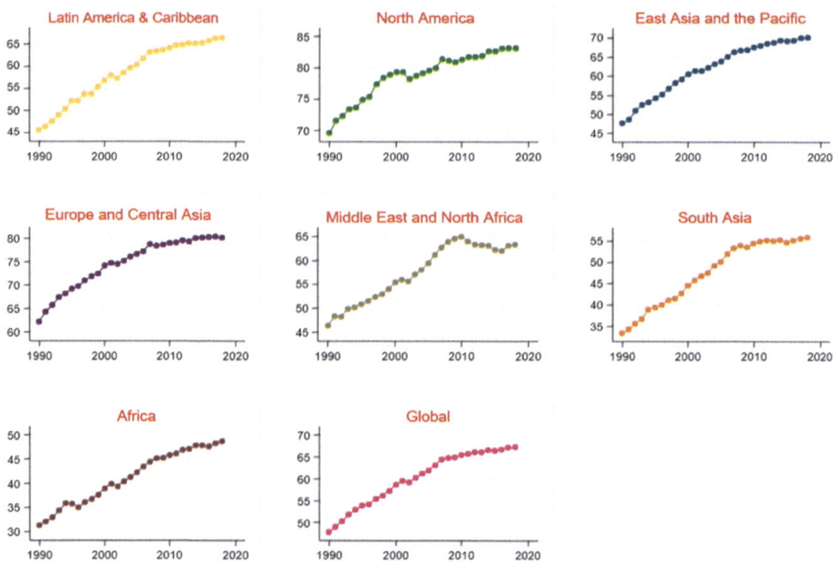

Fig. A2 Evolution of globalization in the regions of the world studied

References

Afshar, M., Sharifi, A., & Mirhashemi, M. (2019). The role of gender differences in assessing the organizational competence of Parsian Bank's management levels. *Social Research, 11*(42), 19–32.

Akinci, M. (2018). Inequality and economic growth: Trickle-down effect revisited. *Development Policy Review, 36*, O1–O24.

Álvarez, M. A. (2020). The confinement of girls and boys in Spain in 2020 by the Covid-19 crisis: Proposals from school social education for the return to school. *Revista De Educación Social, 30*, 457–461.

Antonelli, C., & Tubiana, M. (2020). Income inequality in the knowledge economy. *Structural Change and Economic Dynamics, 55*, 153–164.

Anyanwu, U. M., Anyanwu, A. A., & Cieślik, A. (2021). Does abundant natural resources amplify the negative impact of income inequality on economic growth? *Resources Policy, 74*, 102229.

Appadurai, A. (2003). The production of locality. In *Counterworks* (pp. 208–229). Routledge.

Argadoña, A. (2016). *Inequality is a problem*. IESE Business School.

Asongu, S. A., Orim, S. M. I., & Nting, R. T. (2019). Inequality, information technology and inclusive education in sub-Saharan Africa. *Technological Forecasting and Social Change, 146*, 380–389.

Berger, Z. D., Evans, N. G., Phelan, A. L., & Silverman, R. D. (2020). Covid-19: Control measures must be equitable and inclusive. *British Medical Journal, 368*, 1–2.

Bodemer, K. (1998). Globalization. A concept and its problems. *New society, 156*(6), 54–71.

Bourguignon, F. (2017). The globalization of inequality. *Fondo de Cultura Económica.*

Breunig, R., & Majeed, O. (2020). Inequality, poverty and economic growth. *International Economics, 161*, 83–99.

Canh, N. P., Schinckus, C., Thanh, S. D., & Ling, F. C. H. (2020). Effects of the internet, mobile, and land phones on income inequality and The Kuznets curve: Cross country analysis. *Telecommunications Policy, 44*(10), 102041.

Castelló, A., & Doménech, R. (2002). Human capital inequality and economic growth: some new evidence. *The Economic Journal, 112*(478), C187–C200.

Castelló-Climent, A., & Doménech, R. (2008). Human capital inequality, life expectancy and economic growth. *The Economic Journal, 118*(528), 653–677

Çelik, S., & Basdas, U. (2010). How does globalization affect income inequality? A panel data analysis. *International Advances in Economic Research, 16*, 358–370.

Chenery, H. B. (1975). The structuralist approach to development policy. *The American Economic Review, 65*(2), 310–316.

Chudik, A., & Pesaran, M. H. (2015). Common correlated effects estimation of heterogeneous dynamic panel data models with weakly exogenous regressors. *Journal of Econometrics, 188*(2), 393–420.

Chudik, A., Mohaddes, K., Pesaran, M. H., and Raissi, M. (2016). Long-run effects in large heterogeneous panel data models with cross-sectionally correlated errors. In *Essays in honor of Aman Ullah*. Emerald Group Publishing Limited.

Clouston, S. A. P., Natale, G., & Link, B. G. (2021). Socioeconomic inequalities in the spread of coronavirus-19 in the United States: An examination of the emergence of social inequalities. *Social Science and Medicine, 268* (November 2020), 113554.

Cordera, C. R. (2017). Democracy and the social question: Another turn of the screw. *Economía UNAM, 14*(41), 22–39.

Dumitrescu, E.-I., & Hurlin, C. (2012). Testing for granger non-causality in heterogeneous panels. *Economic Modelling, 29*(4), 1450–1460.

Dussel, E. (2004). World system and transmodernity. *Colonial Modernities*, 201–226.

Elgin, C., & Elveren, A. Y. (2021). Informality, inequality, and feminization of labor. *Women's Studies International Forum, 88*, 102505. https://doi.org/10.1016/j.wsif.2021.102505

Economic Commission for Latin America and the Caribbean [ECLAC]. (2002). *Globalización y desarrollo*. https://www.cepal.org/es/publicaciones/2724-globalizacion-desarrollo
Edwards, S. (1997). Trade policy, growth, and income distribution. *The American Economic Review, 87*(2), 205–210.
Gravina, A. F., & Lanzafame, M. (2021). Finance, globalisation, technology and inequality: Do nonlinearities matter? *Economic Modelling, 96*, 96–110.
Greenwood, D. T., & Holt, R. P. (2010). Growth, inequality and negative trickle down. *Journal of Economic Issues, 44*(2), 403–410.
Gygli, S., Haelg, F., Potrafke, N., & Sturm, J.-E. (2019). The KOF globalisation index revisited. *Review of International Organizations, 14*(3), 543–574 Recovered from: https://doi.org/10.1007/s11558-019-09344-2call_made.
Hall, P. (1998). Globalization and the world cities. In *Globalization and the world of large cities* (pp. 17–36).
Hartwell, C. A., Horvath, R., Horvathova, E., & Popova, O. (2019). Democratic institutions, natural resources, and income inequality. *Comparative Economic Studies, 61*, 531–550.
Huesca, L., & Ochoa, G. (2016). Wage inequality and technological change along the Northern Border of Mexico. *Problemas del Desarrollo, 47*(187), 165–188.
Jauch, S., & Watzka, S. (2016). Financial development and income inequality: a panel data approach. *Empirical Economics, 51*, 291–314.
Kai, H., & Hamori, S. (2009). Globalization, financial depth and inequality in Sub-Saharan Africa. *Economics Bulletin, 29*(3), 2025–2037.
Kino, S., & Kawachi, I. (2020). How much do preventive health behaviors explain education-and income-related inequalities in health? Results of Oxaca-Blinder decomposition analysis. *Annals of Epidemiology, 43*, 44–50.
Kristal, T., & Yaish, M. (2020). Does the coronavirus pandemic level the gender inequality curve? (It doesn't). *Research in Social Stratification and Mobility, 68*(May), 100520.
Kuznets, S. (1955). Economic Growth and Income Inequality. *American Economic Review, 65*, 1–28.
Law, S. H., Naseem, N. A. M., Lau, W. T., & Trinugroho, I. (2020). Can innovation improve income inequality? Evidence from Panel Data. *Economic Systems, 44*(4), 100815.
Le, T. H., Nguyen, C. P., Su, T. D., & Tran-Nam, B. (2020). The Kuznets curve for export diversification and income inequality: Evidence from a global sample. *Economic Analysis and Policy, 65*, 21–39.
Lee, J. E. (2010). Inequality in the globalizing Asia. *Applied Economics, 42*(23), 2975–2984.
Lessmann, C., & Steinkraus, A. (2019). The geography of natural resources, ethnic inequality and civil conflicts. *European Journal of Political Economy, 59*, 33–51.
Lindert, P. H., & Williamson, J. G. (2003). Does globalization make the world more unequal? In *Globalization in historical perspective* (pp. 227–276). University of Chicago Press.
Madsen, J., & Strulik, H. (2020). Technological change and inequality in the very long run. *European Economic Review, 129*, 103532.
Martins-Filho, P. R., Quintans-Júnior, L. J., de Souza Araújo, A. A., Sposato, K. B., Tavares, C. S., Gurgel, R. Q., ... & Santos, V. S. (2021). Socio-economic inequalities and COVID-19 incidence and mortality in Brazilian children: a nationwide register-based study. *Public Health, 190*, 4–6.
McManus, P. A., & Johnson, K. L. (2020). Female labor force participation in the us: How is immigration shaping recent trends? *Social Science Research, 87*, 102398.
Melitz, M. J., & Redding, S. J. (2015). New trade models, new welfare implications. *American Economic Review, 105*(3), 1105–1146.
Mohtadi, S., & Castells-Quintana, D. (2021). The distributional dimension of the resource curse: Commodity price shocks and income inequality. *Structural Change and Economic Dynamics, 59*, 63–78.
Nissanke, M., & Thorbecke, E. (2010). Globalization, poverty, and inequality in Latin America: Findings from case studies. *World Development, 38*(6), 797–802.

Odhiambo, N. M. (2011). Growth, employment and poverty in South Africa. In search of a trickle-down effect. *Journal of Income Distribution, 20*(1), 49–62.
Perugini, C., & Vladisavljevi, M. (2020). Social stability challenged by Covid-19: Pandemics, inequality and policy responses. *Journal of Policy Modeling, 43*(1), 146–160.
Pesaran, M. H. (2004). General diagnostic tests for cross-sectional dependence in panels. *Empirical Economics*, 1–38.
Pesaran, M. H. (2007). A simple panel unit root test in the presence of cross-section dependence. *Journal of Applied Econometrics, 22*(2), 265–312.
Pesaran, M. H. (2015). Testing weak cross-sectional dependence in large panels. *Econometric Reviews, 34*(6–10), 1089–1117.
Pesaran, M. H., & Yamagata, T. (2008). Testing slope homogeneity in large panels. *Journal of Econometrics, 142*(1), 50–93.
Potlogea, A. V. (2018). Globalization and the skilled city. *Journal of Urban Economics, 107*, 1–30.
Ravallion, M. (2020). Ethnic inequality and poverty in Malaysia since May 1969. Part 1: Inequality. *World Development, 134*, 105040.
Reimer, D., Smith, E., Andersen, I. G., & Sortkær, B. (2021). What happens when schools shut down? Investigating inequality in students' reading behavior during Covid-19 in Denmark. *Research in Social Stratification and Mobility, 71*, 100568. https://doi.org/10.1016/j.rssm.2020.100568
Rotunno, L., & Wood, A. (2020). Wage inequality and skill supplies in a globalised world. *Journal of Comparative Economics, 48*(3), 529–547.
Sabi, M. (2007). Globalization and human development. In International conference on globalization and its discontents Cortland, The Sage Colleges, USA.
Sehid M. G., & Ben Ali, M. S. (2021). Globalization and Environmental Pollution: Where Does the MENA Region Stand? In *Economic development in the MENA region.* Pages 161–179.
Sehid M. G., & Ben Ali, M. S. (2023). Do governance indicators predict inequality? A panel non-causality framework for the emerging countries, in *Governance in the developing world transnational insights on economic development* (pp. 43–59).
Solt, F. (2020). Measuring income inequality across countries and over time: The standardized world income inequality database. *Social Science Quarterly, 101*(3), 1183–1199.
Stiglitz, J. (2002). Development policies in a world of globalization. In *New international trends for economic development seminar* (pp. 1–27).
Stiglitz, J. E. (2012). The price of inequality. Allen Lane. https://books.google.com.ec/books?id=k. https://books.google.com.ec/books?id=k_hALgEACAAAAJ.
Tang, S., Wang, Z., Yang, G., & Tang, W. (2020). What are the implications of globalization on sustainability? A comprehensive study. *Sustainability, 12*(8), 3411.
Tsai, P. (1995). Foreign direct investment and income inequality: Further evidence. *World Development, 23*(3), 469–483.
United Nations Development Programme [UNDP]. (2019). *Human Development Report*. https://hdr.undp.org/content/human-development-report-2019
Vega, L. G. (2017). *The Origins Years*. Real editions.
Vera, J, & Poy, S. (2017). Labor market, social policies and inequality: recent changes in historical perspective. Greater Buenos Aires, 1974–2014. *Economía UNAM 14*(42), 3–23.
Vilas, C. M. (1997). State reform as a political issue. *Política y Cultura, 8*, 147–185.
Wan, G. (2007). Understanding regional poverty and inequality trends in China: Methodological issues and empirical lndings. *ReviewiofiIncomeiandiWealth, 53*(1), 25–34.
Wang, C., & Lu, Y. (2020). Can economic structural change and transition explain cross-country differences in innovative activity? *Technological Forecasting and Social Change, 159*(June), 120194.
Westerlund, J. (2007). Testing for error correction in panel data. *Oxford Bulletin of Economics and Statistics, 69*(6), 709–748.
Williamson, J. G. (1965). Regional inequality and the process of national development: a description of the patterns. *Economic Development and Cultural Change, 13*(4, Part 2), 1–84.

Wink, M. V., & Zanandréa, L. H. (2019). Inequality of educational opportunities: Evidence from Brazil. *Economia, 20*(2), 109–120.

World Income Inequality Database [WIID]. (2020). UNU-WIDER. https://doi.org/10.35188/UNU-WIDER/WIID-281123

World Bank. (2020). World Developmenten Indicators. https://datacatalog.worldbank.org/search/dataset/0037712

Zhong, X., Nichols Clark, T., & Sassen, S. (2007). Globalization, producer services and income inequality across US metro areas. *International Review of Sociology—Revue Internationale de Sociologie, 17*(3), 385–391.

Chapter 4
Examining the Long- and Short-Run Asymmetric Effects of Climate Change on Food Security in Tunisia

Amal Jmaii

Abstract Environment change has recently emerged as a prominent global concern, widely attracting the attention of environmentalists. This heightened concern primarily stems from its enduring and detrimental impacts on agricultural and food production, water supply, and, ultimately, household well-being. This chapter examines the relationship between climate change and food security in Tunisia. Using the NARDL model-based cointegration approach, we uncovered significant insights into this relationship. The findings underscore that while mean temperature negatively affects long-term food security, it exerts a positive influence in the short term. Additionally, positive shifts in annual precipitation levels and increased water stress rates notably undermine food security across short- and long-term horizons. Rigorous robustness checks affirm these results. In light of the empirical results, a structured set of critical policy implications is considered. It is strongly recommended that policymakers prioritize future strategies emphasizing adaptation measures and investments in green technologies to mitigate the persistent adverse effects of climate change.

Keywords Food security · Climate change · Asymmetry · NARDL · Water Stress · Tunisia

4.1 Introduction

The Earth's climate is currently undergoing unprecedented changes, progressing at a pace never before observed in the documented climate history. These changes primarily relate to human activities and their associated carbon emissions. Scientific consensus affirms that if these emissions go unmitigated, global average temperature will rise by several degree Celsius by the year 2100. This trend will have severe consequences, leading to significant and far-reaching risks at local, regional, and global levels for both humans and natural ecosystems. Already, the effects of

---

A. Jmaii (✉)
LAREQUAD, FSEGT, University of Tunis El Manar, Manar, Tunisia
e-mail: dramaljmaii@gmail.com

M. S. Ben Ali and E. Lechman (eds.), *Sustainable Economic Development*,
https://doi.org/10.1007/978-981-97-3767-3_4

global climate change are observable and have left their print across all regions of the Earth. Different economic sectors have also been impacted. These effects span on a wide range of areas and include alterations in weather patterns, increased frequency of extreme weather events, rising sea levels, disruptions in ecosystems, threats to biodiversity, and shifts in agricultural productivity, among others. Indeed, one of the primary challenges governments face is ensuring an adequate supply of nutritious food, both in terms of quantity and quality, to meet the increasing demands resulting from population growth and shifting dietary patterns. These demands are escalating as available natural resources continue to diminish. Additionally, it is crucial to consider the impact of climate change on food systems and their ecological foundations, heightening the risks of food insecurity, particularly for the most vulnerable populations.

In response to these challenges, numerous plans were developed and endorsed in 2015, most notably the Sustainable Development Goals (SDGs) and the Paris Agreement. Addressing the challenges posed by climate change, Tunisia has actively participated in developing climate mitigation and adaptation strategies over the past three decades (Soltani and Mellah, 2023). Special attention has been directed toward the agricultural sector, recognizing its pivotal role in ensuring food security. This has involved the implementation of institutional and technical measures aimed at enhancing the resilience of agriculture to climate impacts. Because of its geographical location, Tunisia found itself among the world regions particularly vulnerable to climate variability, notably in terms of erratic precipitation patterns and rising temperatures (Péridy et al., 2012).

Addressing and mitigating the consequences of climate change have become critical to safeguard the planet and its inhabitants.

Tunisia faces escalating challenges because of worsening global warming conditions and dwindling water resources, already below the threshold of absolute water scarcity (Ministry of Agriculture, Water Resources and Fisheries of Tunisia, 2021). These factors pose significant risks that could compound the country's existing economic and social hardships in the forthcoming decades. The projected rise in temperatures, heightened extreme weather events, and a decline in rainfall are anticipated to impact both the quantity and quality of water resources, thus affecting the economy, notably the agricultural sector. Addressing this high water stress will require substantial investments in adaptation measures and enhancing water use efficiency. Given Tunisia's delicate macroeconomic conditions, it becomes imperative to efficiently optimize both public and private resources to reach a sustainable water balance aligned with the country's development objectives (United Nations Climate Change, 2022).

Our primary objective in this study is to assess the effect of climate change indicators—specifically temperature and precipitation—on food security in Tunisia. To this end, we use the NARDL model, to consider the short- and long-run asymmetric effects of these climate change indicators.

The remainder of this paper is structured as follows. Section 4.2 provides an overview of the Tunisian context, highlighting climate conditions and trends in food security. Section 4.3 presents a concise empirical literature review on the impact of climate change on food security. Section 4.4 reports the descriptive statistics of the examined data and outlines the econometric methodology. Section 4.5 details the estimation of the proposed methodology in the Tunisian context. Finally, Sect. 4.6 presents the final findings and concludes with recommendations.

4.2 Tunisian Context: Climate Change and Food Security Landscape in Tunisia

4.2.1 A Highly Variable Climate

The climate of Tunisia, located in North Africa and influenced by the Mediterranean, typically offers pleasant conditions along the coast but turns increasingly extreme as you move south into the desert regions (Fig. 4.1). These climatic conditions have significant implications for agriculture, tourism, and different economic activities in the country. In fact, the Mediterranean Sea exerts a moderating effect on Tunisia's climate, maintaining more moderate temperatures along the coast compared to the interior regions. Specifically, the North of the Tunisian Ridge experiences a Mediterranean climate with a wet season, with the following variations: Wet, Semi-wet, and Sub-humid. In addition, the central area and the Gulf of Gabes[1] experience a semi-arid climate characterized by notable variability, presented as Semi-arid, Sub-arid, and Arid.

Furthermore, in the southern part of the country, particularly in the Desert, Tunisia encounters a desert climate with extremely hot temperatures during summer and colder nights during winter. Officially, the climate in Tunisia is categorized as hot and semi-arid (Fig. 4.1).

On the other hand, the substantial temperature rise in Tunisia since the 1970s (Fig. 4.2) has posed several challenges for the country. In fact, the temperature pattern in Tunisia has been affected by various factors, including climate change and global warming. Over the past few decades, there has been a general upward trend in temperatures, leading to more frequent and intense heat waves during the summer. These heat waves often result in exceptionally high temperatures, sometimes exceeding 40 degree Celsius (104 degree Fahrenheit) in some regions.

Tunisia faces substantial challenges and risks due to worsening global warming conditions and the constraints posed by limited water resources, already below the absolute water scarcity threshold. In fact, the excessive use of groundwater has resulted in a 118% depletion of shallow water tables and a 126% decline in deep water tables. This overexploitation poses an increased risk of salinization and amplifies the economic costs associated with water extraction (United Nations Climate Change, 2022). In fact, the west receives minimal annual precipitation, with a

[1] Located in the southeastern region of Tunisia, Gabes stands as a coastal city and is recognized as the second-largest urban center in southern Tunisia. Its unique feature lies in being both an oasis and a seaport.

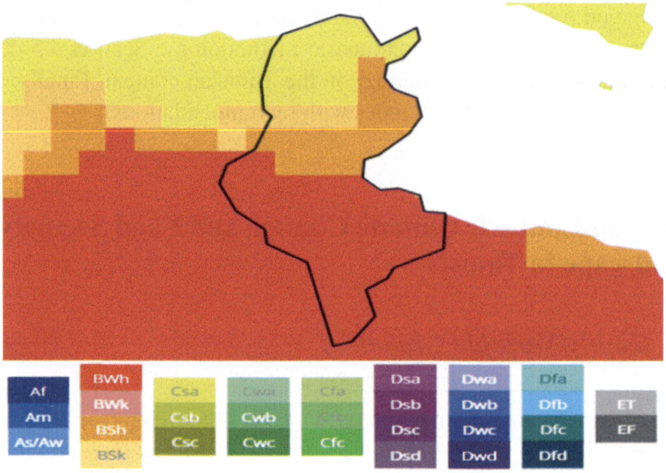

Fig. 4.1 Koppen-Geiger climate classification,[2] Tunisia 1991–2020. *Source* The Climate Change Knowledge Portal CCKP

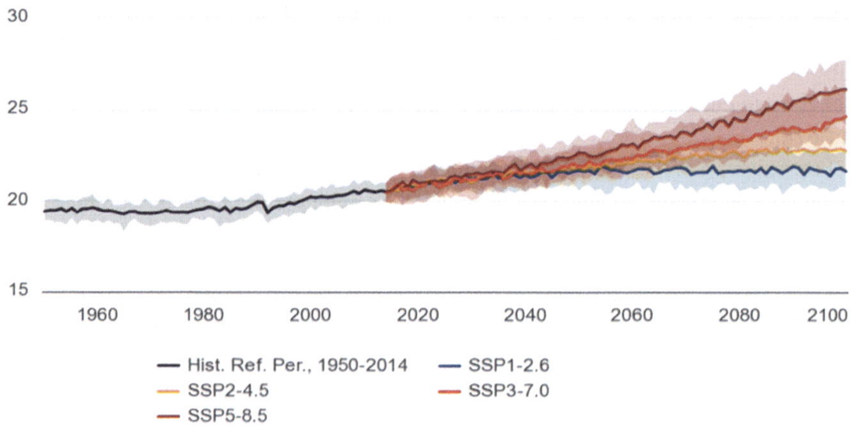

Fig. 4.2 Tunisian projected average mean surface air temperature.[3] *Source* The Climate Change Knowledge Portal CCKP

[2] The Köppen-Geiger climate classification system categorizes climates into five primary groups, and each of these groups is further subdivided based on seasonal patterns of precipitation and temperature (Rubel and Kottek, 2011). These five main groups are designated as A (tropical), B (arid), C (temperate), D (continental), and E (polar). For Tunisia we distinguish two groups, B (arid climate) and C (temperate climate), and four classifications (subgroups): Bwh (Hot desert climate), Bsh (Hot semi-arid climate), Bsk (Cold semi-arid climate), and Csa (Hot-summer Mediterranean climate).

[3] The SSPs, or shared socioeconomic pathways (for further information, refer to O'Neill et al. (2017) pp 1–3), outline different trajectories. SSP1-2.6 aims to remain under 2.5 °C warming, requiring

majority concentrated in coastal regions, while southern areas receive as little as 150 mm per year. Predominant challenges revolve around constrained water resources, changing precipitation patterns, and rising population demands. The country experiences average annual temperatures ranging from 16 °C to 20 °C, with summers marked by hot conditions exceeding 25 °C and reaching maximums of 32 °C. Winters are relatively mild, maintaining average temperatures between 10 °C and 14°C (Tunisia, 2019).

Examination of data retrieved from Fig. 4.2 of the World Bank Group's Climate Change Knowledge Portal (CCKP) shows shared socioeconomic pathway models (O'Neill et al., 2017) used to predict temperature in Tunisia. These models aim to predict how average temperature might change in Tunisia under different scenarios or pathways, including those outlined in the shared socioeconomic pathways (Fig. 4.2). They suggest that without substantial efforts to reduce emissions, temperatures will continue to rise. Degree of warming depends on future emission trajectories and how effective global efforts are in curbing these emissions through mitigation strategies like transitioning to renewable energy sources, improving energy efficiency, implementing sustainable land-use practices, and adopting policies to limit greenhouse gas emissions. However, it is important to note that while projections show a general trend of increasing temperatures, the exact magnitude and regional variations can differ based on complex interactions between various natural and human-influenced factors.

Indeed, one of the most noticeable consequences has been the alteration of the country's climate patterns. This shift has led to prolonged periods of heat waves, erratic rainfall, and altered agricultural cycles, posing challenges to traditional farming practices and affecting crop yields. Farmers have had to adapt their techniques and use new strategies to cope with these changing conditions, sometimes resulting in decreased agricultural productivity.

4.2.2 *Food Production Significantly Contributes to Ensuring Food Security*

Tunisia is expected to experience exacerbated agricultural production losses in the coming decades due to an anticipated rise in temperatures and a reduction in rainfall. Some regions in the country have already borne the severe brunt of climate change, significantly impacting Tunisian agriculture. With approximately 30% arable land, Tunisia has a substantial portion of its territory available for cultivation. This arable land is crucial for the country's agricultural production, including crops like wheat, barley, olives, and citrus fruits. The 27% dedicated to pasture and forests is essential for livestock grazing and maintaining biodiversity and ecosystem services.

net zero emissions in the latter part of the century. SSP2-4.5 projects temperature increases between 2.7 and 3.4 °C by 2100. SSP3-7.0 depicts high emissions without CO_2 and high aerosol emissions. SSP5-8.5 serves as a high reference scenario, devoid of additional climate policies.

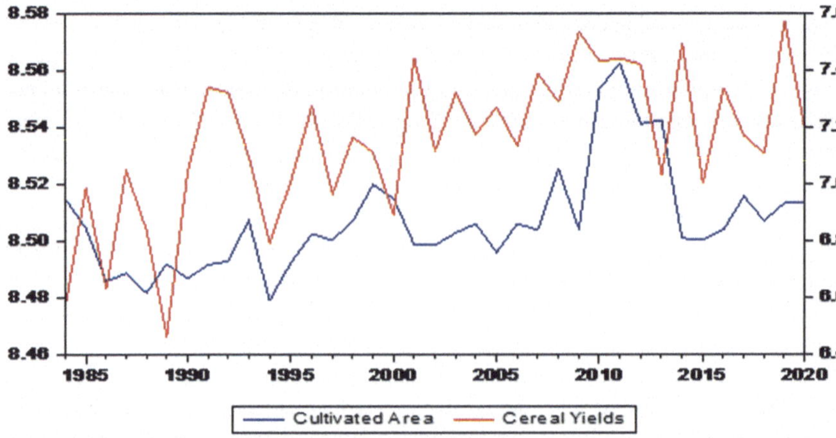

Fig. 4.3 Trend of cereal yields and cultivated area (arable land + permanent crops) in Tunisia. *Source* Author's creation using AQUASTAT data

Forests also provide timber and other non-timber forest products, contributing to the economy. The remaining 43% of agriculturally unusable land typically includes areas like deserts, rocky terrain, and urban development, where agriculture is not feasible. While these areas cannot contribute to traditional agricultural production, they may have other economic uses or environmental importance. However, the agricultural sector, which is crucial for Tunisia's economy, is particularly vulnerable to climate change. Variability in precipitation and rising temperatures can negatively affect crop yields and livestock productivity. The trend of cereal yields and cultivated area (arable land + permanent crops) is shown in Fig. 4.3.

Between 2015 and 2019 (Fig. 4.4), the percentage of people experiencing severe food insecurity in Tunisia increased from 9.1% to 10.7%. This indicates that a larger portion of the population struggles to access an adequate and reliable food supply chain during this period. According to the AFO, in 2021, data shows a significant difference in food insecurity rates between regions of the world. In fact, the percentage of people experiencing severe food insecurity in Africa was estimated at 23.4% in 2021. This is considerably higher than the global average, indicating that a large portion of the African population faces challenges in accessing enough food for a healthy and active life. In contrast, the percentage of people experiencing severe food insecurity in Europe and North America was significantly lower, at only 1.5% in 2021. This suggests that a much smaller proportion of the population in these regions struggles with food access compared to Africa.

These disparities highlight the unequal distribution of resources, infrastructure, and economic opportunities among different parts of the world. Achieving food security and eliminating hunger worldwide is a complex challenge that requires not only addressing immediate food needs but also addressing underlying factors such as poverty, access to education, health care, and economic development. Certainly, climate change has emerged as a critical factor affecting food security in Tunisia, with

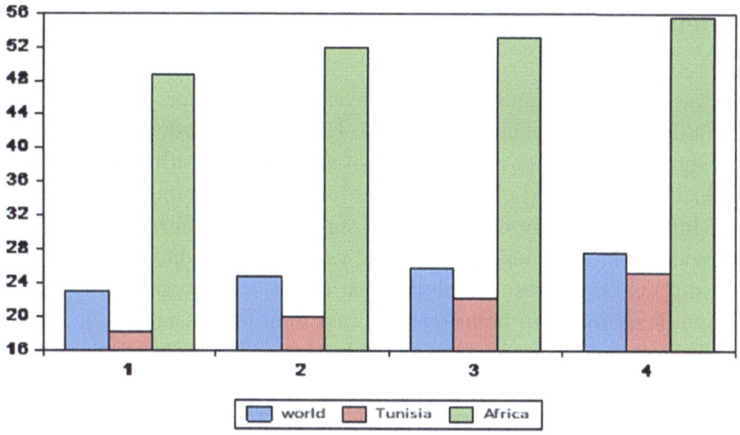

Fig. 4.4 Moderate food insecurity (3-year average). *Source* Author's creation based on FAO data

far-reaching implications for both agricultural production and the well-being of the population. In the Maghreb and North African regions, the year-to-year climate fluctuations have significant and detrimental effects on agriculture, primarily manifested in extended drought periods. These impacts are anticipated to intensify in response to the predicted climate change. Reduction in available water resources will directly affect the agricultural sector, posing a potential threat to Tunisia's socioeconomic development and social stability. This is especially worrying given that rain-fed agriculture serves as the primary occupation and source of livelihood for the vast rural population in the country (Mougou et al., 2011). In fact, rising temperatures and irregular precipitation patterns (Figs. 4.1 and 4.2) linked to climate change have disrupted Tunisia's agricultural sector. Increased temperatures can lead to heat stress in crops and livestock, while changing rainfall patterns can result in prolonged droughts or heavy rainfall, impacting crop yields. Moreover, Tunisia faces water scarcity challenges exacerbated by climate change. Reduced rainfall and increased evaporation rates contribute to dwindling water resources, which are essential for irrigation and agricultural activities. This scarcity affects crop production and overall food availability. In fact, many staple crops in Tunisia, such as wheat, are highly sensitive to temperature increases and water scarcity. Additionally, livestock faces heat stress and reduced animal food availability due to changing climate conditions, impacting meat and dairy production. Furthermore, climate change is causing shifts in optimal agricultural zones, affecting crop suitability and potentially leading to reduced yields in some areas. Farmers may need to adapt by changing the crop types they cultivate (Attiaoui & Boufateh, 2019). Thus, climate-related disruptions in agriculture can lead to food price volatility, making staple foods less affordable for vulnerable populations, especially those in rural areas, where agriculture is a primary source of income and employment. Climate change-related challenges can jeopardize livelihoods in these regions and push households deeper into poverty (Jmaii et al, 2017). This can exacerbate food insecurity and increase malnutrition risk.

4.3 Brief Literature Review

A significant body of research sought to investigate the impact of climate change on the volatility of food security in Tunisia. For instance, Jeder et al. (2020) highlighted a significant long-term causality between food security, measured by the food production index, and climate change indicators (precipitation and temperature). In addition, the study underscored a short-term causal relationship between the food production index and the following variables: land under cereals, inflation, and food imports. These results validate that the issue of food security in Tunisia poses a significant threat in both the short and long terms, indicating a persistent and pressing concern. In a quantitative study, Attiaoui and Boufateh (2019) used an Autoregressive Distributed Lag (ARDL) model to investigate the short-term and long-term effects of climate change on cereal farming in Tunisia. Additionally, they applied Pooled Mean Group (PMG) estimation and conducted Granger causality tests on a regional panel spanning the 1975 to 2014 period. The authors found that climate change issues continue to persist in Tunisia. Furthermore, the study shows that the adverse impacts of climate change have become more pronounced when there is a shortage in rainfall, even though current temperature levels remain conducive to cereal crop growth. Soltani and Mellah (2023) explored the adaptation strategies used by farmers at the farm level in the semi-arid region of Tunisia in response to water shortages. Additionally, they examined the factors that influence and determine the behavior of these farmers in dealing with water scarcity. Examining data from surveys conducted with 81 farmers in the irrigated area of Diar El Hajjaj, a typology of farmers was established. This was achieved through a multivariate approach that integrated clustering methods and a principal component analysis. The results revealed that among the surveyed farmers, there are two distinct groups. These groups tend to use more intensive water resources in their agricultural practices, focusing on cultivating high-value crops and adopting a commercial farming approach.

Moreover, Pickson and Boateng (2022) introduced novel findings on the effect of climate change on food security in Africa. To study climate change trends, the authors applied the Mann–Kendall test and Sen's slope estimator. Moreover, the study used the Pooled Mean Group (PMG) technique and the Dumitrescu-Hurlin panel causality test to examine the impact of climate change on food security in 15 African countries from 1970 to 2016. The findings of the study do not support a robust long-term impact of temperature on food security. However, the short-term results revealed that extreme temperatures can hinder food security, with varying degrees of effects observed across different countries.

In addition, Abbas (2022) investigated the impact of rising average annual temperatures, among other independent variables, on the production of ten key crops in Pakistan, including wheat, rice, bajra, jowar, maize, barley, gram, sugarcane, mustard oil, and cotton. The study covered the 2000 to 2019 period. To understand the nature of both long-term and short-term relationships, the study used different econometric techniques, including panel data unit root tests, the panel PMG model, the panel FMOLS, and the panel DOLS. The long-run results of the study indicate that an increase in annual temperatures has a statistically significant negative effect on the production of major crops. Conversely, the area under cultivation and the use of fertilizers show a statistically significant positive impact on crop production.

4.4 Research Methodology

4.4.1 The Data

The main aim of this section is to scrutinize the asymmetrical causal relationship between climate change, food security, and economic fluctuations in Tunisia. We used time series data spanning from 1987 to 2020, gathered from different sources, including the Penn World Table (PWT), the World Bank, the Food and Agriculture Organization (FAO), AQUASTAT, and the East Anglia (UEA) Climatic Research Unit (CRU) databases, as outlined in Table 4.1.

Table 4.1 Data description

Variables	Definitions	Sources
Gdp	Real GDP on the production side at chained PPPs (in millions of US dollars in 2017) divided by the total population of each country	Penn World Table (PWT)
Inf	Inflation, consumer prices (annual %)	World Bank
Pr	Average annual precipitation (mm)	East Anglia (UEA) Climatic Research Unit (CRU) database
T	Average annual temperature level (°C)	East Anglia (UEA) Climatic Research Unit (CRU) database
FPI	The Food Production Index includes food crops that are not only edible but also contain essential nutrients. However, it excludes coffee and tea from its calculation because, while these beverages are technically edible, they do not have a significant nutritional value	Food and Agriculture Organization (FAO)
WS	Water stress is defined as the ratio between the total freshwater withdrawn by all major sectors and the total renewable freshwater resources, with consideration for environmental water requirements. This measure helps assess pressure on available freshwater resources due to human activities and environmental needs	AQUASTAT/ Food and Agriculture Organization (FAO)

Source Author's creation

In this study, the food production index (FPI) is considered as the dependent variable, while the following independent variables are examined: on the one hand, average annual temperature level (°C), average annual precipitation (mm), and water stress ratio as the main factors behind climate change. On the other hand, we used real GDP and inflation in order to take into consideration economic variability.

Table 4.2 reports the descriptive statistics. In Table 4.2, average values for the variables lFPI, lT, lPR, lINF, lWS, and lGDP are 4.2652, 3.0174, 5.6052, 1.4955, 4.3276, and 7.8456, respectively, while their standard deviations are 0.3117, 0.0234, 0.1866, 0.4067, 0.1341, and 0.4098, respectively.

The correlation matrix presented in Table 4.3 indicates that all variables show a positive correlation with the food production index, except for inflation, which has a negative correlation.

Table 4.2 Descriptive statistics of the variables

	Mean	Median	Sd	Min	Max	Skew	Kurt
LFPI	4.2652	4.3619	0.3117	3.6254	4.8117	−0.2960	2.0979
LGDP	7.8456	7.7776	0.4098	7.1196	8.3680	−0.3102	1.8039
LINF	1.4955	1.4900	0.4067	0.6847	2.1860	−0.0781	2.0559
LPR	5.6052	5.6087	0.1866	5.2735	6.0098	0.0372	2.2369
LT	3.0174	3.0248	0.0234	2.9611	3.0492	−0.9101	3.1744
LWS	4.3276	4.2987	0.1341	4.0319	4.5861	0.1858	2.7840

Source Author's creation
Note sd = standard deviation, Min = minimum, Max = maximum, Skew = Skewness, Kurt = Kurtosis

Table 4.3 Correlation matrix

	lFPI	lGDP	lINF	lPR	lT	lWS
lFPI	1.0000					
lGDP	0.6752	1.0000				
lINF	−0.3136	−0.5029	1.0000			
lPR	0.0830	0.1278	0.0065	1.0000		
lT	0.5345	0.4531	−0.4612	−0.1169	1.0000	
lWS	0.7742	0.4870	0.0347	0.0240	0.5160	1.0000

Source Author's creation

4.4.2 Econometric Methodology

4.4.2.1 NARDL Bound Test for Cointegration

This study uses the recently developed Nonlinear AutoRegressive Distributed Lag (NARDL) technique to examine the asymmetric impact of climate change on food production. Unlike the AutoRegressive Distributed Lag (ARDL) technique, which overlooks nonlinearity and the asymmetric connections between the underlying variables (Chowdhury et al., 2021), the NARDL model, as expanded by Shin et al. (2014), allows us to assess the dynamic adjustment patterns and asymmetric relationships in both the short- and long run between these variables. To delve into the relationship between these variables, we formulate the following model:

$$lFPI = f(lGDP, lINF, lPR, lT, lWS). \tag{4.1}$$

According to Shin et al. (2014), the Nonlinear Autoregressive Distributed Lag (NARDL) model is defined as an asymmetric error correction form, which is specified as follows:

$$\begin{aligned}\Delta lFPI_t &= \alpha_0 + \alpha_1 lFPI_{t-1} + \alpha_2^+ lT_{t-1}^+ \\ &+ \alpha_3^- lT_{t-1}^- + \alpha_4^+ lPR_{t-1}^+ + \alpha_5^- lPR_{t-1}^- + \alpha_6^+ lWS_{t-1}^+ \\ &+ \alpha_7^- lWS_{t-1}^- + \alpha_8 lINF_{t-1} + \alpha_9 lGDP_{t-1} + \sum_{i=1}^{p} \mu_i \Delta lFPI_{t-i} \\ &+ \sum_{m=1}^{p} (\gamma_1^+ \Delta lT_{t-1}^+ + \gamma_2^- \Delta lT_{t-1}^-) + \sum_{m=1}^{p} (\delta_1^+ \Delta lPR_{t-1}^+ \\ &+ \delta_2^- \Delta lPR_{t-1}^-) + \sum_{m=1}^{p} (\theta_1^+ \Delta lWS_{t-1}^+ + \theta_2^- \Delta lWS_{t-1}^-) \\ &+ \sum_{m=1}^{p} \vartheta_1 \Delta lINF_{t-1} + \sum_{m=1}^{p} \vartheta_2 \Delta lGDP_{t-1} + \rho ECT + \varepsilon_{t.}\end{aligned} \tag{4.2}$$

In the given equation, "α_i" represents the long-run coefficients, whereas "μ_i", "γ_i", "δ_i", "θ_i", and "ϑ_i" are the short-run coefficients. These coefficients are used to represent the relationships and effects of various variables in the model over different time horizons, with "α_i" typically associated with long-term effects and the others associated with short-term effects. In addition, the study assumes that the NARDL estimation method is similar to that of the linear ARDL. Therefore, various diagnostic tests are used to assess the appropriateness and robustness of the specified NARDL models.

4.4.3 Checking for Asymmetric Effects

These long-term coefficients quantify the relationship between variables in a state of long-term equilibrium for shocks from the independent variables. By making use of the cumulative dynamic multiplier effect, we can depict the long-term and short-term asymmetric patterns as follows: the impact of a one percent change in V^+ and V^- on Y_t can be expressed through the following equation:

$$e_h^+ = \sum_{i=0}^{h} \frac{\partial lFPI_{t+i}}{\partial lT_t^+};\ e_h^- = \sum_{i=0}^{h} \frac{\partial lFPI_{t+i}}{\partial lT_t^-};\ e_h^+ = \sum_{i=0}^{h} \frac{\partial lFPI_{t+i}}{\partial lPR_t^+};\ e_h^- = \sum_{i=0}^{h} \frac{\partial lFPI_{t+i}}{\partial lPR_t^-};$$
$$e_h^+ = \sum_{i=0}^{h} \frac{\partial lFPI_{t+i}}{\partial lWS_t^+};\ e_h^- = \sum_{i=0}^{h} \frac{\partial lFPI_{t+i}}{\partial lWS_t^-}$$

We assume that when $h \to \infty$ then $e_h^+ \to Le_i^+$ and $e_h^- \to Le_i^-$.

Apart from using the cumulative dynamic multiplier, asymmetry tests can also be used. Specifically, the detection of short- and long-run asymmetry can be achieved through the Wald test. The Wald F-statistics operate under the hypothesis of joint significance, where

$$\begin{cases} H_0 : \rho = \alpha_i^+ = \alpha_i^- = 0 \\ H_1 : \rho \neq \alpha_i^+ \neq \alpha_i^- \neq 0. \end{cases}$$

If the F-statistics exceeds the critical values (Pesaran et al., 2001), this would result in rejecting the null hypothesis, indicating a potential long-run relationship:

$$\begin{cases} H_0 : \rho = 0 \\ H_1 : \rho \neq 0. \end{cases}$$

The hypothesis[4] for the long-run effect is as follows:

$$\begin{cases} H_0 : \rho = \frac{-\alpha_i^+}{\rho} = \frac{-\alpha_i^-}{\rho} = 0 \\ H_1 : \rho \neq \frac{-\alpha_i^+}{\rho} \neq \frac{-\alpha_i^-}{\rho} \neq 0. \end{cases}$$

[4] The same hypotheses apply for the short run; we only replace the coefficient for the short run ("μ_i", "γ_i", "δ_i", "θ_i", and "ϑ_i").

Table 4.4 Unit root test

variables	Augmented Dickey-Fuller (ADF)		Phillips-Perron (pp)		Decision
	Level	First	Level	First Difference	
lfpi	−0.4860 (0.8798)	−4.1721*** (0.0031)	−1.6711 (0.4369)	−28.2239*** (0.0001)	I(1)
Lgdp	−2.1979 (0.2108)	−4.6973*** (0.0007)	−2.1687 (0.2209)	−4.6896*** (0.0007)	I(1)
Linf	−2.3452 (0.1648)	−7.8205*** (0.0000)	−2.3087 (0.1754)	−7.8884*** (0.0000)	I(1)
IT	−3.2727** (0.0248)	–	−3.2727** (0.0248)	–	I(0)
IPR	−6.6314*** (0.0000)	–	−15.4226*** (0.0000)	–	I(0)
LWS	−1.1725 (0.6756)	−4.3457*** (0.0015)	−1.3595 (0.5909)	−4.4270*** (0.0012)	I(1)

Source Author's creation

4.5 Results and Economic Interpretation

4.5.1 Stationarity

The next step in this study involves assessing the stationarity of the underlying variables to ensure that none of them has an integration order I(2). This is crucial because the Nonlinear AutoRegressive Distributed Lag (NARDL) model requires that the variables be integrated at the order I(0) or I(1) to investigate cointegration among them. To determine integration order, unit root tests were conducted, specifically, the augmented Dickey-Fuller (ADF) and Phillips-Perron (PP) tests. The results are summarized in Table 4.4.

The results of the unit root tests indicate that mean temperature, average annual precipitation, and the water stress ratio show an integration order I(0), indicating that they are stationary and do not possess a unit root. Conversely, domestic production growth, inflation rate, and the food production index show stationarity after undergoing the first differencing (I(1)).

4.5.2 Short- and Long-Run Asymmetric Effect of Climate Change on Food Security

The results of the proposed methodology reveal intriguing insights. First of all, both positive and negative fluctuations in mean temperature have adverse and statistically significant effects on the food production index. *Ceteris paribus*, when average temperature increases by 1%, it leads to a significant reduction in food production by approximately 13.8% in the long run. This suggests that any deviation, whether it is an increase or decrease in mean temperature, has a detrimental impact on food production. In fact, when there are unexpected increases in mean temperature, it has a harmful impact on food production. Higher temperatures can lead to a number of problems in crop cultivation, including heat stress on plants, reduced photosynthesis, and changes in the timing of growth stages, all of which can ultimately lead to decreased crop yields, and, therefore, food production. On the other hand, negative shocks on mean temperature can lead to slower growth, delayed maturation, and increased vulnerability to diseases and pests, all of which can result in reduced rice harvests. Moreover, the study highlights the negative impact of GDP on food production in the short term. This finding fits perfectly the Tunisian context. In fact, since the 2011 uprising, sometimes economic growth may take a while to translate into increased investment in the agricultural sector. In the short term, this delay in investment can lead to production constraints and hindered food output. Another reason that may explain this result is that economic growth can sometimes contribute to environmental degradation, such as deforestation and pollution, which can negatively impact agricultural productivity in the short term.

Additionally, Table 4.5 shows that a negative deviation in water scarcity has a more pronounced and lasting impact in the long term. This implies that when water scarcity improves or decreases (i.e., negative deviation), the consequences and effects are more significant and enduring over an extended period. This finding highlights the importance of addressing and mitigating water scarcity issues, especially in long-term planning and sustainable resource management (Table 4.6).

4.5.3 Asymmetry Study and Diagnostic Tests

The asymmetry test confirms the asymmetric effects of the selected climate change indicators. In addition to the asymmetry tests, we plot the following cumulative dynamic multiplier graphs. We introduced multiple dynamic adaptations, as outlined in Fig. 4.5, illustrating the cumulative dynamic multipliers. These multipliers unveil the evolution of food price production over time, as it moves toward a new long-term equilibrium following positive or negative unitary shocks in precipitations, mean surface temperature, and water stress. Computed through the Akaike information criterion (AIC) best-fit NARDL model, these dynamic multipliers effectively demonstrate adjustment patterns. The distinct positive (solid black line) and negative (black dotted line) curves outline food production's responsive trajectory to these shocks, offering a vivid portrayal of the adjustments.

The cumulative dynamic multiplier graph confirms the negative correlation with the food production index. This aligns with the findings presented in Table 4.5, indicating that a negative deviation, compared to a positive deviation, in water scarcity for example, exerts a stronger influence over the long term. In addition, Fig. 4.5 confirms the inverse relationship between precipitation and the food price index. Negative

Table 4.5 Short-term and long-term asymmetric effects of climate change

Variables	Coefficient	Std. Error	Prob.
Short term			
D(LFPI(t-1))	−0.1999**	0.0708	0.0135
D(LGDP)	−0.8682***	0.1862	0.0004
D(LINF)	0.9375	1.4468	0.5455
D(LT (+))	−10.3199***	1.5821	0.0005
D(LT (−))	−9.5865***	1.0708	0.0000
D(LPR (+))	25.1340***	4.6386	0.0003
D(LPR (−))	−14.0725**	5.4508	0.0273
D(LWS (+))	9.711***	4.8120	0.0001
D(LWS (−))	−11.1956***	4.5114	0.0001
ECT(t-1)	−0.9110***	0.1105	0.0000
Long term			
Constant	54.1480***	12.8876	0.0085
LGDP	−9.6634	11.9647	0.4281
LINF	15.9312***	2.6195	0.0017
LT (+)	87.4454	12.7617	0.1347
LT (−)	−13.7476***	7.3198	0.0023
LPR (+)	−6.8628	6.4119	0.3334
LPR (-)	8.9566**	7.2554	0.0104
LWS (+)	23.8812***	6.0884	0.0055
LWS (−)	−37.7934**	6.1832	0.0389
F-Bounds test	10.7520		
I(1) I(0)	2.85 1.85		

Note(s) The superscripts *, **, and *** denote significance at the 1, 5, and 10% levels
Source Author's creation

Table 4.6 Asymmetry tests

	Short run		Long run		Joint (short and long run)	
	F-statistic	probability	F-statistic	probability	F-statistic	probability
LT	3.8921	0.0625	11.3397	0.0031	5.7467	0.0106
LPR	4.2973	0.0513	3.2377	0.0871	6.4324	0.0069
LWS	3.0480	0.0810	5.2390	0.0222	2.6196	0.0731

Source Author's creation

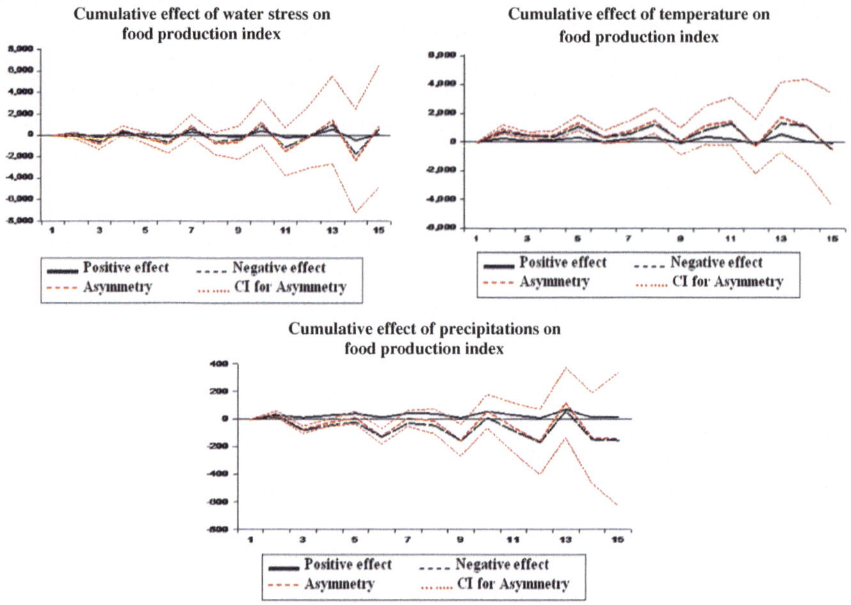

Fig. 4.5 Dynamic multiplier adjustment graph. *Source* Author's creation

changes in precipitation have a more pronounced effect than positive changes over the forecast period. Additionally, there is a notable asymmetric response to precipitation fluctuations. Moreover, a significant asymmetric response is evident regarding shocks in mean surface temperature. Similar to precipitation, the food production index demonstrates a negative correlation. This validation concurs with the findings detailed in Table 4.5, underscoring the greater long-term impact of a negative shock in mean temperature compared to a positive shock.

Table 4.7 presents the results of various diagnostic tests conducted to assess the reliability of the model, including assessments for normality, autocorrelation, and heteroscedasticity. Notably, the NARDL model shows no diagnostic issues, indicating its robustness. To further assess the stability of the model, CUSUM and CUSUMQ tests were conducted. As shown in Fig. 4.6, the predicted line falls within the critical values at the 5% significance level, affirming the model's high degree of stability (Fig. 4.6).

4.5.4 Robustness Check

In order to check the robustness of our results, we used an ARDL-based quadratic specification. The interest in using the quadratic model lies in determining the nonlinear relationship between several specified variables within the model, using

Table 4.7 Results of diagnostic tests

Diagnostic tests	Statistics
Jarque–Bera (*Normality*)	1.6033 (0.4485)
Breusch-Godfrey LM (*Serial correlation*)	0.2849 (0.7550)
ARCH (*Heteroskedasticity*)	0.8884 (0.3537)

(.) are p values
Source Author's creation

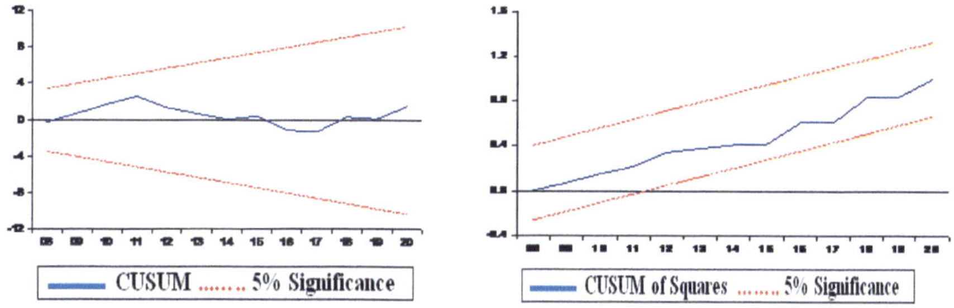

Fig. 4.6 CUSUM and CUSUM of Squares tests. *Source* Author's creation

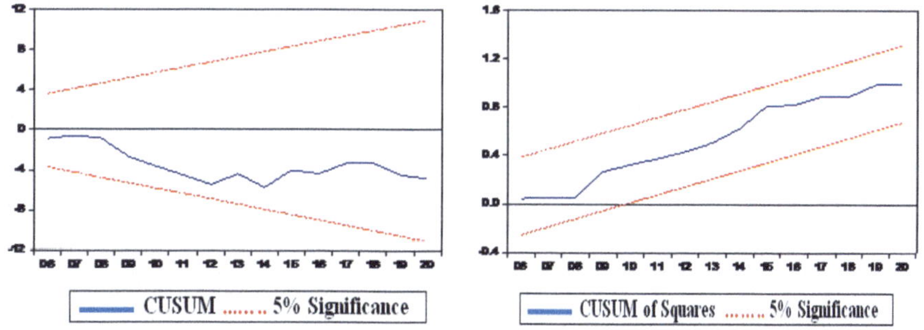

Fig. 4.7 CUSUM and CUSUM of Squares tests of the quadratic model. *Source:* Author's creation

econometric specifications and based on an asymptotic distribution of the variables of interest. The methodology aims to estimate quadratic equations that can be represented using a "U"-shaped curve, indicating a minimum, or an inverted "U", indicating a maximum. Indeed, the relationship between the threshold of climate change indicators and the level of the food production index can be illustrated using a second-degree polynomial function.

Therefore, we incorporate quadratic terms for temperature, precipitation, and water stress in order to address potential nonlinearities and the consequent threshold impacts of climate change on food production. Following Wang (2012), Alagidede et al. (2016), and Li et al. (2019), we propose the following quadratic specification:

$$\begin{aligned} LFPI_t &= \alpha_0 + \beta_1 LGDP_t + \beta_2 LINF_t + \beta_3 LT_t \\ &\quad + \beta_4 LT_t^2 + \beta_5 LPR_t + \beta_6 LPR_t^2 + \beta_7 LWS_t + \beta_8 LWS_t^2 + \varepsilon_t. \quad (3) \end{aligned}$$

The results in Table 4.8 confirm the nonlinear relationship between the different climate change indicators and the food production index. In fact, analysis of the relationship between the food production index and climate change indicators, using a quadratic specification, revealed that, overall, the signs of the coefficients remained consistent across all models. This underscores the resilience and reliability of the proposed methodology.

The diagnostic tests presented in Table 4.8 showcase the robustness of the selected model. Furthermore, to enhance our evaluation of the model's stability, CUSUM and CUSUMQ tests were used. In Fig. 4.7, the predicted line remains within the critical values at the 5% significance level, confirming the model's robust stability.

4.6 Conclusion and Policy Recommendations

The aim of the present study is to examine the asymmetric connection between climate change and food security in Tunisia. This study used a nonlinear AutoRegressive Distributed Lag model, to estimate time series data spanning from 1985 to 2020. The study's findings underscore the presence of asymmetric relationships between the selected variables over both short- and long-run periods. This suggests that the impacts of climate change on food production in Tunisia are not uniform and can vary depending on specific circumstances and time horizons. Therefore, understanding these asymmetries is crucial for developing effective strategies to ensure food security and agricultural sustainability in the face of a changing climate. First, given the observed asymmetrical relationship between climate change and food production, policymakers should prioritize the development and promotion of climate-resilient agricultural practices. This might include the adoption of drought-tolerant crop varieties, improved irrigation techniques, and sustainable soil management practices. Second, recognizing the significance of water scarcity, policymakers should focus on efficient water resource management. Investments in water infrastructure, water conservation measures, and sustainable water use practices can help mitigate the negative impacts of water scarcity on agriculture and food security.

To effectively implement necessary adaptation measures addressing climate change, the Tunisian government must establish requisite legal frameworks and regulations. While efforts have been made, a critical missing component is an absence

Table 4.8 short-term and long-term asymmetrical effects of climate change.

Variables	Coefficient	Std. Error	Prob.
Short term			
D(LFPI(t-1))	−0.2709**	0.1176	0.0289
D(LGDP)	−45.9531**	21.5216	0.0416
D(LINF)	−0.1151***	0.0569	0.0523
D(LT)	−46.6597**	18.3065	0.0162
D(LT^2)	3.6508*	0.6623	0.0000
D(LPR)	−36.2698*	10.6497	0.0020
D(LPR^2)	7.61E-06*	7.96E-07	0.0000
D(LWS)	−1046.563*	89.5555	0.0000
D(LWS^2)	–	–	–
ECT(t-1)	−0.9548***	0.1683	0.0000
Long term			
Constant	16.4654	128.8916	0.8996
LGDP	0.6199*	0.1261	0.0001
LINF	−0.2888**	0.1078	0.0141
LT	4.9437	51.9292	0.9251
LT^2	−0.0127	0.0644	0.8452
LPR	−1.0066***	0.5273	0.0700
LPR^2	8.69E-09**	3.40E-06	0.0185
LWS	−5.4249*	1.5069	0.0017
LWS^2	0.0005*	0.0001	0.0004
F-Bounds test	10.2344		
I(1) I(0)	2.11 3.15		
Diagnostic tests			
Jarque–Bera (Normality) 0.1125 Breusch-Godfrey LM (Serial correlation) 0.3261 ARCH (Heteroskedasticity) 0.5821			

Note(s) The superscripts *, **, and *** denote significance at the 1, 5, and 10% levels.
Source Author's creation.

or a lack of enhanced laws and regulations, specifically targeting environmental and climate-related issues. The legislation serves as a pivotal tool in addressing climate change by offering a legal framework for implementing and enforcing policies to reduce greenhouse gas emissions promote sustainability, and safeguard the environment. Consequently, the country needs comprehensive climate change legislation outlining its commitment to mitigating and adapting to climate change. This legislation can furnish a legal basis for climate action, delineating clear targets and responsibilities across different sectors. Furthermore, ensuring effective enforcement of these legal frameworks and regulations is imperative. This may require penalties for non-compliance and mechanisms enabling citizens to seek redress in cases of climate-related harm.

Appendix

We consider that $lT_t^+, lT_t^-; lPR_t^+, lPR_t^-; lWS_t^+, lWS_{t-1}^-$ are partial sum processes of positive and negative variations in, $lT, lPR, andlWS$, respectively (Baig et al, 2022), with

$$lT_t^+ = \sum_{i=1}^{t} \Delta lT_i^+ = \sum_{i=1}^{t} \max(\Delta lT_i, 0)$$

$$lT_t^- = \sum_{i=1}^{t} \Delta lT_i^- = \sum_{i=1}^{t} \min(\Delta lT_i, 0)$$

$$lPR_t^+ = \sum_{i=1}^{t} \Delta lPR_i^+ = \sum_{i=1}^{t} \max(\Delta lPR_i, 0)$$

$$lPR_t^- = \sum_{i=1}^{t} \Delta lPR_i^- = \sum_{i=1}^{t} \min(\Delta lPR_i, 0)$$

$$lWS_t^+ = \sum_{i=1}^{t} \Delta lWS_i^+ = \sum_{i=1}^{t} \max(\Delta lWS_i, 0)$$

$$lWS_t^- = \sum_{i=1}^{t} \Delta lWS_i^- = \sum_{i=1}^{t} \min(\Delta lWS_i, 0).$$

References

Alagidede, P., Adu, G., & Frimpong, P. B. (2016). The effect of climate change on economic growth: Evidence from Sub-Saharan Africa. *Environmental Economics and Policy Studies, 18*, 417–436. https://doi.org/10.1007/s10018-015-0116-3.

Abbas, S. (2022). Climate change and major crop production: Evidence from Pakistan. *Environmental Science and Pollution Research, 29*(4), 5406–5414. https://doi.org/10.1007/s11356-021-16041-4.

Attiaoui, I., & Boufateh, T. (2019). Impacts of climate change on cereal farming in Tunisia: a panel ARDL–PMG approach. *Environmental Science and Pollution Research, 26*, 13334–13345. https://doi.org/10.1007/s11356-019-04867-y.

Baig, I. A., Chandio, A. A., Ozturk, I., Kumar, P., Khan, Z. A., & Salam, M. A. (2022). Assessing the long-and short-run asymmetrical effects of climate change on rice production: empirical evidence from India. *Environmental Science and Pollution Research*, 1–22. https://doi.org/10.1007/s11356-021-18014-z.

Chowdhury, M. A. F., Meo, M. S., Uddin, A., & Haque, M. M. (2021). Asymmetric effect of energy price on commodity price: New evidence from NARDL and time frequency wavelet approaches. *Energy, 231*, 120934. https://doi.org/10.1016/j.energy.2021.120934.

Food and Agriculture Organization (FAO). *World Food and Agriculture—Statistical Yearbook (2021)*. 1–368. https://doi.org/10.4060/cb4477en.

Jeder, H., Hattab, S., & Frija, I. (2020). An econometric analysis for food security in Tunisia. *New Medit:Mediterranean Journal of Economics, Agriculture and Environment= Revue Méditerranéenned'Economie Agriculture et Environment, 19*(4). https://doi.org/10.30682/nm2004a.

Jmaii, A., Rousselière, D., & Daniel, C. (2017). Semi-parametric Regression-based Decomposition Methods: Evidence from Regional Inequality in Tunisia. *African Development Review, 29*(4), 660–673. https://doi/abs/10.1111/1467-8268.12304.

Li, N., Bai, K., Zhang, Z., Feng, J., Chen, X., & Liu, L. (2019). The nonlinear relationship between temperature changes and economic development for individual provinces in China. *Theoretical and Applied Climatology, 137*, 2477–2486. https://doi.org/10.1007/s00704-018-2744-6.

Ministry of Agriculture, Water Resources and Fisheries of Tunisia (2021). Rapport national sur le secteur de l'eau.http://www.onagri.tn/uploads/secteur-eau/RNE2021_VersionFianle_MarsF2023_ONAGRI.pdf.

Mougou, R., Mansour, M., Iglesias, A., Chebbi, R. Z., & Battaglini, A. (2011). Climate change and agricultural vulnerability: a case study of rain-fed wheat in Kairouan, Central Tunisia. *Regional Environmental Change, 11*, 137–142. https://doi.org/10.1007/s10113-010-0179-4.

O'Neill, B. C., Kriegler, E., Ebi, K. L., Kemp-Benedict, E., Riahi, K., Rothman, D. S., & Solecki, W. (2017). The roads ahead: Narratives for shared socioeconomic pathways describing world futures in the 21st century. *Global Environmental Change, 42*, 169–180. https://doi.org/10.1016/j.gloenvcha.2015.01.004.

Péridy, N., Ghoneim, A., & Brunetto, M. (2012). *The economic costs of climate change in MENA countries: A micro-spatial quantitative assessment and a survey of policies*. Femise Research Programme 2010–2011, FEMISE Project n°34-03.

Pesaran, M. H., Shin, Y., & Smith, R. J. (2001). Bounds testing approaches to the analysis of level relationships. *Journal of Applied Econometrics, 16*(3), 289–326. https://doi.org/10.1002/jae.616

Pickson, R. B., & Boateng, E. (2022). Climate change: a friend or foe to food security in Africa?. *Environment, Development and Sustainability*, 1–26. https://doi.org/10.1007/s10668-021-01621-8.

Rubel, F., & Kottek, M. (2011). Comments on:" The thermal zones of the Earth" by Wladimir Köppen (1884). *Meteorologische Zeitschrift, 20*(3), 361. https://doi.org/10.1127/0941-2948/2011/0258

Shin, Y., Yu, B., & Greenwood-Nimmo, M. (2014). Modelling asymmetric cointegration and dynamic multipliers in a nonlinear ARDL framework. *Festschrift in honor of Peter Schmidt: Econometric methods and applications*, 281–314. https://doi.org/10.1007/978-1-4899-8008-3_9.

Soltani, L., & Mellah, T. (2023). Exploring farmers' adaptation strategies to water shortage under climate change in the Tunisian semi-arid region. *Environmental Management, 71*(1), 74–86. https://doi.org/10.1007/s00267-022-01604-z.

Tunisia (2019). Tunisia's Third National Communication as part of the UNFCCC. URL: https://unfccc.int/sites/default/files/resource/Synth%C3%A8se%20Ang%20Finalis%C3%A9.pdf

United Nations Climate Change (2022). Stratègie de dèveloppement neutre en carbonne et rèsilient au changement climatique à l'horizon 2050-Tunisie. United Nations Climate Change). https://unfccc.int/sites/default/files/resource/Strat%C3%A9gie%20de%20d%C3%A9veloppement%20neutre%20en%20carbone%20et%20r%C3%A9silient%20-%20Tunisie.pdf.

Wang, K. M. (2012). Modelling the nonlinear relationship between CO2 emissions from oil and economic growth. *Economic Modelling, 29*(5), 1537–1547. https://doi.org/10.1016/j.econmod.2012.05.001

Chapter 5
Neoliberalism, Climate Risks, and Resilience-Building in the Caribbean

Shirley-Ann Behravesh and Jonas Gamso

Abstract This study examines the relationship between the adoption of neoliberal economic policies and climate change-related risks in the Caribbean. These neoliberal policies, which promote free trade, foreign investment inflows, and austerity measures, have steered Caribbean economies towards a heavy reliance on tourism—a reliance that has, in turn, heightened the region's vulnerability to certain climate risks, such as extreme storms and rising sea levels. These risks threaten the tourism industry, as well as the local livelihoods that depend on it. This chapter investigates this topic in greater detail. In doing so, it offers empirical evidence that neoliberal policies have led to an expansion of the tourism sector in Caribbean countries. It then considers how this expansion of tourism has affected the climate vulnerabilities of these nations. The chapter closes by suggesting strategies that can be used to enhance resilience to climate change, emphasizing the need for alignment with the UN Sustainable Development Goals through a multifaceted approach that involves policymakers, businesses, and community groups.

5.1 Introduction

The Caribbean region faces significant economic and social challenges (Cannonier & Burke, 2022; Mohammadi et al., 2022; Witter, 2021), which are intricately linked to the broader objectives of the UN Sustainable Development Goals (UNSDG). These challenges stem, in part, from the legacies of neoliberal, "Washington Consensus" policies adopted by Caribbean countries in the late twentieth and early twenty-first Centuries. These policies, which were adopted at the behest of Washington-based institutions such as the International Monetary Fund (IMF) and the US Treasury in response to the debt crisis that had swept the region in the 1980s (see Klack, 1997),

S.-A. Behravesh · J. Gamso (✉)
Thunderbird School of Global Management, Arizona State University, Phoenix, AZ, USA
e-mail: jonas.gamso@thunderbird.asu.edu

S.-A. Behravesh
e-mail: satab2@thunderbird.asu.edu

M. S. Ben Ali and E. Lechman (eds.), *Sustainable Economic Development*,
https://doi.org/10.1007/978-981-97-3767-3_5

opened up Caribbean economies to free trade and FDI inflows while encouraging governments to cut spending and privatize state-owned enterprises (Williamson, 2003).

The consequences of Washington Consensus policies are complex and multifaceted, but one significant impact has been the expansion of the Caribbean tourism sector relative to traditional sectors such as agriculture (Peterson, 2023; Witter, 2021). This is concerning against the backdrop of climate change, as the tourism industry in the Caribbean's coastal and small island nations is especially vulnerable to extreme storms and rising sea levels, which have been exacerbated by rising global temperatures. These impacts of changing weather patterns threaten to destroy key tourism-related infrastructure and deter travelers from visiting (Jerez Columbié, 2022; Pathak et al., 2021). To the extent that Caribbean countries have become dependent on tourism and on the tourism industry, this vulnerability puts their national economies and the livelihoods of many of their residents at risk.

We explore these matters in greater depth in the pages that follow. In doing so, we investigate how neoliberal policies have shifted the economies of Caribbean countries toward tourism and how this shift is affecting nations' climate vulnerabilities. From there, we discuss the potential pathways by which Caribbean countries may enhance their resilience to climate change, highlighting the roles and responsibilities of policymakers, businesses, and civil society groups.

This takes on profound importance considering the international community's commitment to sustainable development and to the UNSDGs, particularly goals related to economic growth (Goal 8), sustainable cities and communities (Goal 11), and reduced inequalities (Goal 10), though connections to several other goals exist (Jerez Columbié, 2022). The impact of neoliberal policies on this region potentially undermines sustainable economic growth by increasing vulnerabilities to external shocks that stand to exacerbate inequalities within societies. Addressing this challenge requires a nuanced understanding of past policy impacts and a concerted effort to move toward policy alignment with the UNSDGs. The experiences of the Caribbean region underscore the importance of adopting a holistic approach to development that integrates closely with sustainability principles.

5.2 Background and Literature Review

The Caribbean region consists of the Caribbean Sea and its islands, as well as some nearby coastal areas on the mainland. The Caribbean Sea, which covers about 2.7 million square kilometers, is separated from the Pacific Ocean by Mexico and Central America. The islands of the Caribbean are divided into three major groups: the Greater Antilles, the Lesser Antilles, and the Lucayan Archipelago. The Greater Antilles includes the four large islands of Cuba, Jamaica, Hispaniola (shared by Haiti and the Dominican Republic), and Puerto Rico. The Lesser Antilles are further subdivided into the Leeward Islands, the Windward Islands, and the Leeward Antilles. The Lucayan Archipelago comprises the Bahamas and the Turks and Caicos Islands.

The Caribbean region also includes some parts of the mainland that have cultural and historical ties with the islands, including Belize, Guyana, Suriname, French Guiana, and some coastal areas of Central and South America.

5.2.1 Economic Development in the Caribbean

Historically, the economies of the region were predominantly based on agriculture. This can be traced back to the colonial era when European countries brought in enslaved and indentured workers from West Africa, India, and China. These laborers were employed in plantations that produced cotton, sugarcane, and other tropical crops (Morgan, 2019). The Caribbean economies were historically characterized by rampant monoculture, such that the economic structure of individual countries often relied heavily on a single crop (Chawla, 2004). While this economic model was ingrained from the onset of colonial rule, it persisted even after most Caribbean countries gained independence in the 1960s.

Caribbean governments, seeking to diversify and industrialize their economies after achieving independence, adopted import substitution industrialization (ISI) policies in the 1960s and 1970s. The idea, grounded in structuralist trade theory, was that these countries should import inputs to drive their home-grown manufacturing sectors, as opposed to importing manufactured goods from abroad (Bulmer-Thomas, 2012; Gafar, 1979). While the ISI strategy carried some benefits for Caribbean countries, such as stimulating the growth of domestic industries and generating employment, the industrial policies that supported it were quite costly, driving much of the region into deep debt (Girvan et al., 1991). Debt service costs became unsustainable as an economic crisis struck the region in the 1970s and 1980s (Hilaire, 2001), forcing Caribbean countries to seek loans from international financial institutions (IFIs)—particularly the IMF.

Loans from IFIs were accompanied by economic policy conditions known as stabilization and structural adjustment programs, which required borrowers to adopt Washington Consensus policies (Girvan et al., 1991)—so named because they were widely promoted by Washington-based institutions, such as the IMF, the World Bank, and the US Treasury. These policies were derived from the assumption that the free market and minimal state intervention were the best ways to stabilize the economy, control inflation, and integrate the region into the global economy. They emphasized the importance of private enterprise, openness to trade, and foreign investment, and advocated for a series of reforms to liberalize, privatize, and deregulate the domestic economies of Caribbean countries. They also urged governments to maintain fiscal discipline, competitive exchange rates, and sound monetary policies to avoid macroeconomic imbalances and crises (Babb & Kentikelenis, 2021; Spence, 2021).

Washington Consensus policies were influenced by the neoliberal ideology that dominated the political and intellectual discourse in the West at the time (Harvey, 2007). While Washington Consensus policies largely followed what economists considered to be best practices (Spence, 2021), they have been criticized for offering one-size-fits-all solutions (Stiglitz, 2008), fostering inequality (Forster et al., 2019), leading to a weakening of labor and environmental standards (Broad & Cavanagh, 1999), and undermining climate action (Fremstad & Paul, 2022) among other things (Rodrik, 2006).

5.2.2 Tourism and Climate Vulnerability

In the Caribbean context, the adoption of Washington Consensus policies contributed to the shift to a tourism-based economy as traditional output and export sectors waned (IMF, 2019; Witter, 2021). The region's agricultural sector could not adequately respond to the rapidly growing demand for high-standard, agri-food products from tourism, processing, and retailing sectors inside and outside the region. Instead, the growing demand in these regional sectors was mainly fulfilled by imports, which crowded out local farmers (CDB, 2019). In contrast, the tourism industry in the Caribbean flourished into the twenty-first century, as air travel became economically accessible to middle-class North Americans and Europeans and as large-scale sea cruises to the region became commonplace (Baker, 2022).

While tourism has been an important source of economic activity, its centrality to the region's economic well-being has proven risky. In particular, any shocks that stop the flow of tourists or destroy key infrastructure, such as hotels, ports of entry, or natural attractions, can lead to economic collapse, as communities that depend heavily on tourism have little to no capacity to respond to the loss of their primary revenue source.

St. Lucia exemplifies the challenges facing many Caribbean economies. Until the 1980s, agriculture was the country's dominant industry. During that decade, St. Lucia embarked on a comprehensive development program to diversify its economy and penetrate foreign markets. Leveraging substantial public investment and foreign aid, the island aimed for steady growth by broadening its key sectors, notably tourism and manufacturing (Meditz & Hanratty, 1987). Yet only tourism boomed and today it accounts for approximately 65% of the island's economic activity, while other sectors have stagnated. Consequently, St. Lucia's economy is vulnerable to shocks like intense storms, global recessions, terrorism, and pandemics, as evidenced by a 7% tourism decline after 9/11 and significant infrastructural damage from Hurricane Tomás in 2010 (Lewis, 2022; National Hurricane Center, n.d.). The COVID-19 pandemic further stressed these vulnerabilities, bringing tourism to a halt and leading to an estimated 8.5% GDP contraction in 2020 (US Department of State, n.d.).

As this example demonstrates, the implications of dependence on a vulnerable industry are significant. This is especially true given the growing threats posed by climate change. In particular, rising global temperatures have contributed to an increase in the intensity of extreme storms (AghaKouchak et al., 2020; Clarke et al., 2022), as well as to rising sea levels (Edwards et al., 2021). The limited geographic and economic scale of Caribbean islands means that disruptions from extreme climate-related weather events, such as hurricanes, can devastate large portions of local economies and cause widespread damage to crops, water supplies, infrastructure, and other critical resources and services, as well as the hotels, beaches, and air and sea ports that support tourism. Rising sea levels pose a similar threat, as key tourism-supporting infrastructure may become submerged. The economic costs of climate change inaction in the Caribbean are projected to total US$10.7 billion per year by 2025, largely reflecting the loss of tourism revenue (FCDO, 2021).

For example, the Bahamas, predominantly recognized for its beach tourism, is likely to face substantial repercussions due to increasing global temperatures and the accompanying rise in sea levels. A one-meter increase in sea level would result in the submersion of 80% of the Bahamas, thereby categorizing it among the Caribbean islands most vulnerable to sea-level rise and storm surge (CCCRA, 2012; Mackay & Spencer, 2017; Pathak et al., 2021). Current models for sea-level rise predict the inundation of 60% of their beaches, which would dramatically affect the available hotel room count on the island (CCCRA, 2012; Pathak et al., 2021). Consequently, the future of these islands is precarious, with their survival intricately linked to climate change.

Other Caribbean nations face similar threats. In 2017, in a span of two weeks, two Category 5 hurricanes, Irma and Maria, caused widespread devastation in the Caribbean. Scientists attribute the extreme nature of these storms, in part, to climate change (Drash, 2017). The destruction was immense, with Antigua & Barbuda, Saint Martin, and Sint Maarten experiencing over 85% damage due to Irma (PAHO, 2017). Maria left Dominica with over 90% of its infrastructure destroyed, critically damaging water, electricity, and communication services (PAHO, 2017). These hurricanes also significantly impacted tourism, as American visitor numbers dropped by 79% in Sint Maarten and 45.6% in Puerto Rico compared to the previous year (Carballo Chanfón et al., 2023). This highlights the substantial threat that extreme weather, exacerbated by global warming, poses to Caribbean tourism and to the economies that rely on it.

5.3 Hypotheses

The literature discussed above suggests that Washington Consensus policies have led Caribbean countries to depend on tourism and that this dependence has, in turn, contributed to heightened climate vulnerabilities. However, these assertions must be scrutinized empirically. Therefore, we investigate hypotheses corresponding to each:

Hypothesis #1 Washington Consensus policies are positively associated with the growth of the tourism industry in the Caribbean

Hypothesis #2 The growth of the Caribbean tourism industry is positively correlated with climate vulnerabilities in Caribbean countries

5.4 Research Methods

We test these hypotheses using a country-level dataset that covers 21 Caribbean and Caribbean adjacent nations from 1998–2020.[1] First, we conduct multiple regressions to determine whether the implementation of Washington Consensus policies has been accompanied by an expansion of the tourism industry in Caribbean countries. After presenting the results of our regression analyses, we conduct a series of correlation tests to assess whether tourism-oriented economies are more vulnerable to climate risks. Variables[2] and estimation models are discussed in the paragraphs that follow.

5.4.1 Variables

The dependent variable in our regression models is tourism receipts which we draw from the World Bank's World Development Indicators (WDIs). This measure captures "expenditures by international inbound visitors, including payments to national carriers for international transport." We divide tourism receipts by total economic activity, measured with per capita gross domestic product (GDP), in order to capture the significance of tourism to economic well-being.

The independent variables are measures that capture Washington Consensus policies and the cross-border economic flows that they produce. We first account for tariff rates. Trade liberalization is among the key tenets of the Washington Consensus, and so it is appropriate to account for tariff rates in our models. Moreover, tariff rates serve as a proxy for neoliberal policies more generally, as countries that adopt this facet of the Washington Consensus are likely to adhere to its other policy recommendations. We also account for FDI inflows and exports (each logged), as these are the sorts of economic flows that are most associated with Washington Consensus policies. Data for these variables is drawn from the WDIs.

We control for several variables, using data from the WDIs, in order to isolate the relationship between the Washington Consensus and tourism receipts. First, we control for per capita income, on the basis that wealthier nations will have more robust and diversified economies, leading a variety of sectors to benefit from heightened investment and trade flows. Second, we control government effectiveness, as evidence suggests that sound political institutions foster greater economic diversification (Matallah, 2020). Lastly, we account for internet users as a share of the population, which serves as a proxy for industrial capacity. We expect more industrialized countries to have more diversified economies.

[1] The countries in our analysis are Antigua and Barbuda, the Bahamas, Barbados, Dominica, Dominican Republic, Grenada, Haiti, Jamaica, St. Kitts and Nevis, St. Lucia, Trinidad and Tobago, Belize, Colombia, Costa Rica, Guatemala, Guyana, Honduras, Mexico, Nicaragua, Panama, and Suriname.

[2] Descriptive statistics for all variables included in this study are presented in Table 5.2 in the Appendix.

5.4.2 Estimation Methods

We analyze this data using linear regression models for panel data. We include country and year-fixed effects to account for unobserved factors across time and place, as well as robust standard errors, clustered by country, to account for heteroscedasticity and autocorrelation. Independent and control variables are lagged one year to give them time to affect the dependent variable.

5.4.3 Assessing Climate Risk Implications

After using regressions to evaluate the relationships between neoliberal policies and the expansion of the tourism industry, we conduct additional analysis to determine whether tourism-oriented Caribbean nations are relatively vulnerable to the impacts of climate change. In doing so, we evaluate three indicators of climate risk. First, we assess the *Climate Vulnerability Index* from the Notre Dame Global Adaptation Initiative, which measures a country's exposure, sensitivity, and ability to adapt to the negative impact of climate change, with attention to food, water, health, ecosystem service, human habitat, and infrastructure.[3] Next, we consider two specific measures of climate vulnerability that are especially relevant to Caribbean nations facing extreme storms and rising sea levels: *Water Stress*[4] and the number of people exposed to natural disasters such as hurricanes.[5]

5.5 Empirical Results and Implications for Climate Risks

As hypothesized, our results suggest that neoliberal policies have enhanced the tourism industries in Caribbean economies. As Table 5.1 shows, the coefficient for *Tariff Rate* is negative and significant, meaning that tourism decreases as tariffs rise. This indicates that neoliberal policies to lower tariffs have led to a more robust tourism sector. Likewise, *Exports* have a significant and positive relationship with *Tourism Receipts*, suggesting that higher levels of trade correspond to a larger tourism industry. The coefficient for *FDI Inflows* is not significant and the direction of the coefficient changes from model to model, such that we cannot say that a relationship exists here.

While these results suggest that neoliberal policies have increased Caribbean countries' reliance on tourism, an important caveat is that trade and trade policy seem

[3] This index ranges from 0 to 1, where greater climate vulnerability corresponds to a higher score.

[4] This variable, which is drawn from the WDIs, measures country-level "freshwater withdrawal as a proportion of available freshwater resources."

[5] This variable comes from the EM-DAT database, from the Center for Research on the Epidemiology of Disasters (CRED). We log the variable to account for its heavy skew.

Table 5.1 Linear panel data models

Dependent variable: Tourism receipts	Model 1	Model 2	Model 3	Model 4
Tariff rate	−7,683*			−9,448**
	(3,951)			(3,971)
FDI inflows (logged)		7,572		−1,974
		(6,717)		(6,664)
Exports (logged)			132,701**	172,755**
			(49,835)	(60,734)
GDPpc (logged)	−52,567	−30,060	−77,604	−151,432
	(120,740)	(99,981)	(136,544)	(153,928)
Government effectiveness	49,330*	31,193	22,997	11,737
	(27,209)	(30,193)	(35,163)	(45,289)
Internet users (% population)	−296.2	−318.1	730.7	545.9
	(960.3)	(773.4)	(950.6)	(1,150)
Observations	246	317	241	182
R-squared	0.464	0.392	0.519	0.616
Number of countries	21	21	17	16
Country and year fixed effects	Yes	Yes	Yes	Yes

Robust standard errors in parentheses, clustered by country; independent and control variables lagged one year; *** $p < 0.01$, ** $p < 0.05$, * $p < 0.1$

to be having more of an impact than FDI. This is likely reflective of the challenges that the Caribbean agricultural sectors faced when they were no longer protected by tariffs. As noted above, these sectors struggled to compete against foreign producers, whereas tourism proved resilient to open trade policies. In contrast, FDI is often targeted toward new industries, promoting economic diversification in the process (Longmore et al., 2014).

5.5.1 Implications for Climate Risks

Having established that there is a positive relationship between neoliberal, Washington Consensus policies and the importance of Caribbean countries' tourism industries, we now reflect on what this means for the climate change-related challenges facing these nations. To do so, we assess relationships between the robustness of the tourism industry and climate vulnerability. This exercise paints a mixed picture, with some climate indicators suggesting that dependency on tourism is accompanied by greater environmental strain and another showing evidence that tourism-focused economies are less environmentally vulnerable. Of course, these results should be taken with caution, as correlation does not necessarily correspond to causality, yet we hope they will be a starting point for future research.

We look, first, at the *Climate Vulnerability Index*, to which tourism receipts have a negative correlation ($r = -0.1927$). This relationship is visualized in Fig. 5.1. On its face, this suggests that more tourism corresponds to lower climate risk. Indeed, the growth of tourism can lead to better infrastructure (Blake et al., 2006), such as improved roads, airports, and coastal defenses that enhance a country's ability to withstand and respond to climate-related events like hurricanes and sea-level rise. A thriving tourism industry can also heighten awareness of the need to protect natural resources, which are often key attractions for tourists (Chen et al., 2023). This may lead to greater funding and efforts toward environmental conservation and climate change mitigation strategies.

However, the vulnerability index measure, which is often used in the climate risk literature (e.g., Chen et al., 2022), has the limitation of bundling together several variables. Thus, if tourism is leading to a worsening of one dimension of climate risk and the betterment of another, the index may show a neutral outcome. Even as the relative growth of the tourism industry exacerbates water stress, for example, it may also lead to a reduction in climate vulnerabilities associated with other sectors, such as agriculture (Parker et al., 2019).

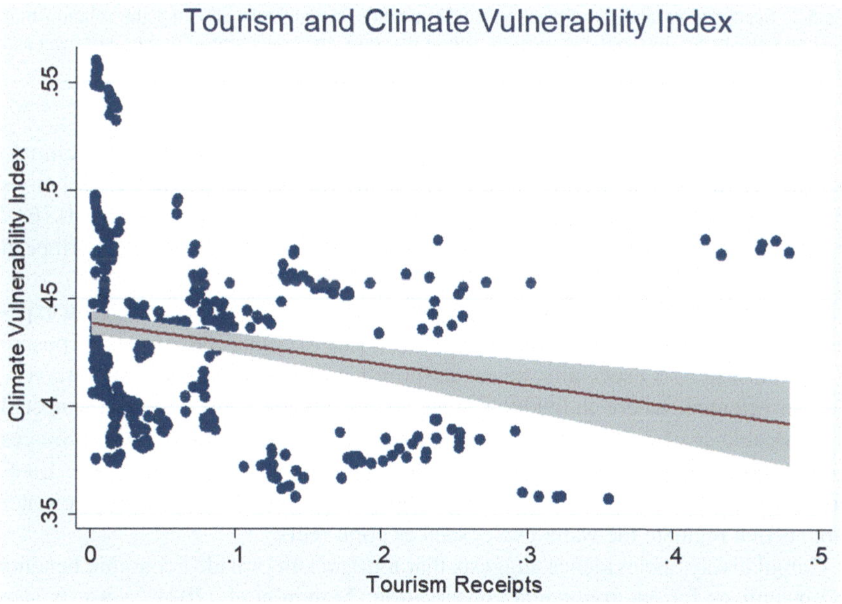

Fig. 5.1 Scatter plot and trendline, with 95% confidence intervals, showing the relationship between tourism receipts and the Climate Vulnerability Index for 21 Caribbean and Caribbean adjacent countries from 1998–2020

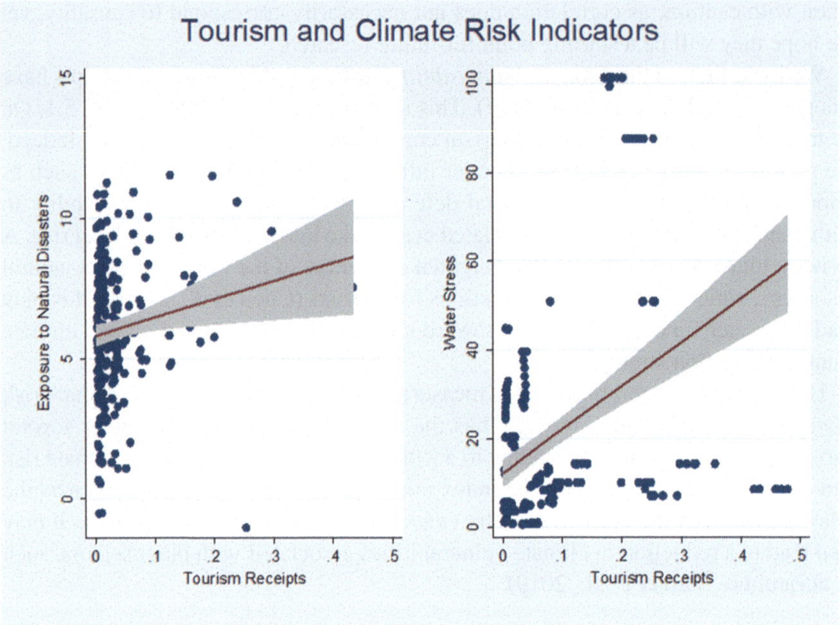

Fig. 5.2 Scatter plots and trendlines, with 95% confidence intervals, showing the relationships between tourism receipts and exposure to natural disasters and water stress for 21 Caribbean and Caribbean adjacent countries from 1998–2020

Therefore, we consider alternative measures that more directly measure climate risks of relevance to Caribbean nations. In doing so, we find positive correlations between the tourism receipts variable and *Exposure to Natural Disasters* ($r = 0.163$), as well as between tourism receipts and the level of water stress facing Caribbean countries ($r = 0.3945$). These relationships are visualized in Fig. 5.2.

These latter results are in line with Hypothesis #2. Caribbean countries are experiencing heightened exposure to extreme storms due to climate change and exposure to these disasters is especially severe in those Caribbean economies that are very tourism oriented, where individuals' lives and livelihoods are directly connected to vulnerable coastal areas. Additionally, tourism can compromise water resources through pollution generated by sewage, chemicals, and other contaminants. Likewise, the activities of tourists, such as boating, diving, and fishing, can damage water habitats that regulate the water cycle, such as coral reefs.

Cumulatively, the evidence suggests that tourism can provide economic benefits and incentives for environmental conservation (Mahon et al., 2021), yet it is also leading to some significant environmental risks (Liu et al., 2022). Balancing these factors is crucial for Caribbean countries to effectively manage their exposure to climate risks. As we discuss in the next section, stakeholders spanning sectors can play important roles in enhancing climate resilience in the Caribbean.

5.6 Strategies for Building Climate Resilience

A comprehensive strategy that encompasses not only environmental considerations, but also socio-economic factors is essential for managing the climate-related threats to increasingly tourism-dominated Caribbean nations. This approach requires the active participation of various stakeholders, including national governments, intergovernmental organizations, members of the private sector, and local communities, and alignment with sustainability principles such as the UNSDGs. Each of these entities brings valuable resources, perspectives, and expertise, such that they can each contribute to building climate resilience through their own efforts. Yet it is also essential that they coordinate and collaborate to help Caribbean nations thrive in a rapidly changing environment.

Conceptually, we frame this in top-down, bottom-up terms, such that policymakers at the national and supranational levels can offer high-level policy reforms to address climate risks, whereas members of the private and civil society sectors can offer solutions that more directly impact society. It is also important to note that private sector and civil society actors are stakeholders that can and should have a voice in shaping policy changes from above.

From above and below, these parties can play a crucial role in aiding small, tourism-oriented countries to manage their climate vulnerabilities. Additionally, they can help move Caribbean nations away from the Washington Consensus policies that have left them exposed to climate risks. That does not mean abandoning trade and investment openness, which can carry significant benefits in terms of advancing economic growth (Raghutla, 2020) as well as environmental policy betterment (Gamso, 2017), but it does mean finding ways to protect Caribbean economies and societies from the accompanying risks.

5.6.1 Top-Down Approaches: Climate Risk Management and Policy Adjustment

We first focus on efforts that policymakers can take to foster greater climate resilience. National governments can do so by investing in sustainable infrastructure that can withstand extreme weather events, promoting eco-tourism, diversifying their economies through innovation, and promoting inclusive and sustainable economic growth. They can also implement environmental regulations to protect their natural resources, which are often the main attraction for tourists.

Indeed, many Caribbean states are working to establish national adaptation plans (Kemp-Benedict et al., 2020; Lewis & Su, 2021; Thomas et al., 2019). However, these plans remain underdeveloped in terms of laying out measures for implementation and monitoring progress (Woodruff & Regan, 2019), calling into question how readily they can be activated in practice (Thomas et al., 2020). More generally, developing countries often have difficulty corralling resources (Besley & Persson, 2014) and carrying out large-scale sustainability projects (Sarvajayakesavalu, 2015), raising concerns about their adaptive capacity when it comes to managing climate risks (Eriksen et al., 2021).

This creates an avenue for members of the international community to offer assistance, whether in the form of funding, expertise, or policy guidance. For example, foreign donors can provide financial aid and technical support for projects that build climate resilience, such as seawalls, renewable energy installations, and sustainable water management systems. Indeed, the United States, the European Union, and the World Bank, among other multilateral and bilateral donors, are increasingly providing climate-oriented foreign aid to low- and middle-income countries (Weiler et al., 2018) and the Caribbean region has received hundreds of millions of dollars in climate and disaster relief funding (Roy, 2023). Yet the sustainability financing gap persists, both worldwide (United Nations Environment Programme, 2023) and in Caribbean nations specifically (Mohan, 2023). Thus, there is a need for more foreign assistance to support Caribbean countries' climate change management efforts.[6]

Members of the international community can also help to advocate for climate action on behalf of Caribbean countries, which are heavily affected by climate change, even as they themselves have done relatively little to contribute to it. Intergovernmental organizations can amplify Caribbean voices in global forums, pushing for stronger commitments to reduce greenhouse gas emissions and for the implementation of the Paris Agreement—including urging developed nations to follow through on their financing pledges (Peake & Ekins, 2017).

Efforts can also be made to encourage the diffusion of best practices across countries. Regional bodies, such as the Caribbean Community (CARICOM), the Association of Caribbean States (ACS), and the Organisation of Eastern Caribbean States (OECS), may be particularly well situated to coordinate these efforts, as they have a shared understanding of the common challenges and vulnerabilities that Caribbean countries face due to climate change, and can foster a sense of solidarity and cooperation among them.

While efforts to adapt Caribbean countries to the impacts of climate change are essential, these countries' governments should also take steps to reshape their economies by moving them away from strict neoliberal models, which often emphasize minimal government intervention and free-market dominance. The goal should be to adopt strategies that retain the beneficial aspects of neoliberalism, such as attracting opportunities for trade and foreign investment, while addressing its shortcomings that have made Caribbean economies vulnerable to the impacts of climate change. To this end, key policy interventions should include government subsidies and programs in support of emerging sectors like technology and renewable energy, as well as entrepreneurship. Investing in education and skill development is another crucial aspect. This approach will make Caribbean workforces more attractive for

[6] It should be noted that there is considerable debate about the effectiveness of foreign assistance (Qian, 2015). While it is beyond the scope of this chapter to review the debate in detail, it is worth mentioning that the effectiveness of aid is likely conditional on the strategic interests of donors (Bearce & Tirone, 2010), the sectors that aid is targeted towards (Jones & Tarp, 2016), and the institutional environment in aid-recipient countries (Gamso, 2023).

high-value, technologically advanced industries. Policy efforts such as these can help to diversify Caribbean economies, thereby limiting the climate risks wrought by an overdependence on tourism and on the tourism industry.

That said, diversifying a country's economic base is financially demanding and requires strong political will and efficient governance. These are areas in which Caribbean countries often struggle and that have been strained under neoliberal regimes of limited government. This offers regional and intergovernmental organizations a role to play in facilitating a shift toward more sustainable and equitable economic models. Financial and technical assistance from these actors can help Caribbean countries overcome the challenges of moving away from neoliberal policies. Even as they work toward more balanced and diversified economies, these organizations can help maintain and enhance trade and investment opportunities, ensuring that Caribbean countries do not lose the benefits of neoliberalism's open economic engagement.

Crucially, many of these bodies, including the IMF and the World Bank, have often championed neoliberal principles, promoting policies like market liberalization, privatization, and deregulation. Yet, if they are to help Caribbean nations transition from the Washington Consensus, they will need to undertake a shift in mindset and policy, moving away from strict neoliberal constraints and toward a more resilient and inclusive economic future. While there had been considerable optimism about a shift along these lines in the years after the global financial crisis, recent evidence suggests that IFIs have reverted back to a policy approach more consistent with the Washington Consensus (Güven, 2018), such that a sustained shift in mindset is more necessary now than ever, particularly given the consensus among both critics and champions of neoliberalism that these policies have not achieved their desired results (Rodrik, 2006).

5.6.2 *Bottom-Up Approaches: Community Innovation and Influencing Policy*

While policy efforts from above are clearly important, policy reforms thus far have been quite limited. Despite advancements in climate science, a lack of political will among national governments has led to policy stagnation. Likewise, international institutions have had minimal impact on reducing net losses and vulnerability to climate change on both the global and national scales. However, many local communities have independently focused on their own climate-related problems, often with little publicity (Gabriel-Campos et al., 2021; Rasmussen, 2014; Schreuder & Horlings, 2022), decreasing their vulnerabilities to extreme weather events and climate change (Schlingmann et al., 2021; Schreuder & Horlings, 2022). In doing so, they have grown their economies and nurtured civic pride, among other community values. Not every community has succeeded, but successful communities serve as working models of reducing or adapting to climate change, which other communities can build upon.

Citizen-led initiatives are crucial in fostering climate resilience in economically challenged communities worldwide (Zapata Campos et al., 2023). These initiatives are centered around the mobilization of local resources, establishing governance structures, and enhancing local investment capabilities, thereby empowering communities. In particular, grassroots innovations (GIs) are a bottom-up approach primarily led by local communities or non-profit organizations. Over the last two decades, GIs have emerged as a useful approach to building sustainable communities and adapting to the impacts of climate change (Belda-Miquel et al., 2020; Hossain, 2016; Patnaik & Bhowmick, 2020). These innovations are likely to be more sustainable in tackling local problems at the community level as they benefit from local perspectives and knowledge.

Some examples of such innovations include those emerging from alternative energy communities, the repair movement, food networks, eco-housing, community currencies, and the transition town movement (Zapata Campos et al., 2023). These movements and communities are characterized by the multiple organizational forms in which they manifest themselves—such as cooperatives, voluntary associations, community organizations, and companies—supported by different resource bases, motivations, and contexts (Hargreaves et al., 2013).

Civil society groups and activists are also pivotal in advocating for climate change policy adoption. These groups raise awareness about climate change at local and national levels and assist governments, donors, and international organizations in planning for a climate-constrained future (Shi & Moser, 2021). Civil society networks engage in a wide range of activities, including mobilizing public opinion and supporting climate change adaptation activities that benefit the most vulnerable, developing low-carbon development pathways, conducting research, disseminating results, building capacity on climate change, and influencing government planning processes at various levels through advocacy activities.

These advocacy activities often target the poor and vulnerable, who are most affected by climate change due to their location, climate-sensitive livelihoods, and limited resources. Lobbying activities can include arranging small face-to-face meetings with decision-makers or sending letters or useful reports or information to these decision-makers. Campaigning activities can include educating people about an issue, arranging public meetings, writing newsletters, asking people to sign petitions, preaching a sermon, or taking part in demonstrations or marches. Most civil society advocacy activities target national-level government policies, but a significant amount of advocacy occurs at the international level too. Many advocates contribute to this through the Climate Action Network—a worldwide network of over 700 NGOs in more than 90 countries that works to promote government and individual action to limit human-induced climate change to ecologically sustainable levels—and through other channels. Civil society networks in vulnerable countries also focus some of their advocacy efforts on donors, such as the World Bank, which may be active in their countries, as well as the private sector (Reid et al., 2012; Shi & Moser, 2021).

Businesses, as part of civil society, can also influence government planning processes at a multitude of levels through a variety of advocacy activities. For example, by adopting sustainable practices, they can demonstrate the feasibility and benefits of such practices, thereby influencing policy. Businesses can also lobby for policies that support sustainable practices. Likewise, individuals, through their daily choices and political actions, can significantly influence policy and greenhouse gas (GHG) emissions. Among other things, by choosing environmentally friendly products and services, individuals can drive demand for sustainable practices in businesses, thereby indirectly influencing policy (Dietz et al., 2020).

Finally, the media plays a crucial role in shaping public opinion and influencing policy. Formal and informal media sources reach many who may not be fully aware of the issues or their ability to be part of the solution. More recently, forms of social media have played an important role in raising awareness of climate change, highlighting risks, and building individual agency through sharing possible solutions. By accurately reporting on the impacts of climate change and the need for adaptive policies, the media can raise public awareness and pressure governments to act (Reid et al., 2012; Schäfer & Painter, 2021).

5.6.3 Integrating Top-Down and Bottom-Up Approaches

Integrating top-down and bottom-up strategies is necessary for more effective adaptation to climate change. This approach acknowledges the potential of adaptation activities to simultaneously address vulnerability to climatic and non-climatic stressors. Integrating top-down and bottom-up approaches benefits from shared resources, information, and aspirations, and the collaboration of multiple stakeholders may help address deficiencies of current adaptation strategies. Moreover, collaboration at multiple levels helps enable adaptive co-management systems, which allow for dynamic and flexible problem-solving and responses to change (Amaru & Chhetri, 2013).

5.7 Conclusions

As natural disasters increase in frequency and intensity due to rising global temperatures, climate risks have become a significant area of concern for Caribbean nations. The region's vulnerability to extreme storms and rising sea levels has been exacerbated by its increasing reliance on tourism, which was driven in large part by the adoption of neoliberal, Washington Consensus policies. This chapter has explored these matters empirically, while offering pathways by which national governments, intergovernmental organizations, private sector organizations, and civil society actors can support economic diversification and climate resilience, in alignment with the UNSDGs, including the goals on sustainable economic growth (Goal 8) and sustainable cities and communities (Goal 9).

Future studies should aim to deepen our understanding of the nuanced interplay between economic development, tourism dependency, and climate resilience in the Caribbean. By expanding the scope of research in these areas, we can better comprehend the challenges faced by the Caribbean and develop more effective strategies to mitigate climate risk while promoting sustainable economic growth.

Appendix

See Table 5.2.

Table 5.2 Descriptive statistics for all variables used in this study

Variable	Mean	Std. Dev.	Minimum	Maximum
Tourism receipts	0.0680173	0.0876617	0.0020417	0.4791812
Tariff rate	10.66235	5.0417	2.52	33.36
FDI inflows (logged)	19.64248	1.936202	13.85464	24.65367
Exports (logged)	22.3471	1.697882	18.7883	26.92306
Per capita GDP (logged)	9.387639	0.6029626	7.962108	10.53753
Government effectiveness	−0.056361	0.7032357	−2.170481	1.563377
Internet users (% Pop.)	27.75563	25.37179	0	95.70763
Climate Vul. index	0.4355798	0.0428256	0.3573588	0.5600128
People Aff. by disasters	6.183745	2.47714	−1.026042	11.56336
Water stress	15.99553	21.55442	0.7221703	101.25

References

AghaKouchak, A., Chiang, F., Huning, L. S., Love, C. A., Mallakpour, I., Mazdiyasni, O., Moftakhari, H., Papalexiou, S. M., Ragno, E., & Sadegh, M. (2020). Climate extremes and compound hazards in a warming world. *Annual Review of Earth and Planetary Sciences, 48*, 519–548.

Amaru, S., & Chhetri, N. B. (2013). Climate adaptation: Institutional response to environmental constraints, and the need for increased flexibility, participation, and integration of approaches. *Applied Geography, 39*, 128–139.

Babb, S., & Kentikelenis, A. (2021). Markets everywhere: The Washington Consensus and the sociology of global institutional change. *Annual Review of Sociology, 47*, 521–541.

Baker, D. M. A. (2022). Caribbean tourism development, sustainability, and impacts. In C. Cannonier & M. G. Burke (Eds.), *Contemporary issues within Caribbean economies* (pp. 235–264). Palgrave Macmillan.

Bearce, D. H., & Tirone, D. C. (2010). Foreign aid effectiveness and the strategic goals of donors. *The Journal of Politics, 72*(3), 837–851.

Belda-Miquel, S., Pellicer-Sifres, V., & Boni, A. (2020). Exploring the contribution of grassroots innovations to justice: Using the capability approach to normatively address bottom-up sustainable transitions practices. *Sustainability, 12*(9), 3617.

Besley, T., & Persson, T. (2014). Why do developing countries tax so little? *Journal of Economic Perspectives, 28*(4), 99–120.

Blake, A., Sinclair, M. T., & Campos Soria, J. A. (2006). Tourism productivity: Evidence from the United Kingdom. *Annals of Tourism Research, 33*(4), 1099–1120.

Broad, R., & Cavanagh, J. (1999). The death of the Washington Consensus? *World Policy Journal, 16*(3), 79–88.

Bulmer-Thomas, V. (2012). *The economic history of the Caribbean since the Napoleonic wars.* Cambridge University Press.

CARIBSAVE Climate Risk Atlas [CCCRA]. (2012). *Climate change risk profile for The Bahamas.* https://www.caribank.org/publications-and-resources/resource-library/thematic-papers/study-state-agriculture-caribbean

Cannonier, C., & Burke, M. G. (2022). *Contemporary issues within Caribbean economies.* Springer International Publishing.

Carballo Chanfón, P., Mohan, P., Strobl, E., & Tveit, T. (2023). The impact of hurricane strikes on cruise ship and airplane tourist arrivals in the Caribbean. *Tourism Economics, 29*(1), 68–91.

Caribbean Development Bank [CDB]. (2019). *Study on the state of agriculture in the Caribbean.* https://www.caribank.org/publications-and-resources/resource-library/thematic-papers/study-state-agriculture-caribbean

Chawla, R. L. (2004). Caribbean economies: Dependency and beyond. *India Quarterly*, 218–242.

Chen, J., Huang, Y., Wu, E. Q., Ip, R., & Wang, K. (2023). How does rural tourism experience affect green consumption in terms of memorable rural-based tourism experiences, connectedness to nature, and environmental awareness? *Journal of Hospitality and Tourism Management, 54*, 166–177.

Chen, Y., Zhang, D., Wu, F., & Ji, Q. (2022). Climate risks and foreign direct investment in developing countries: The role of national governance. *Sustainability Science, 17*, 1723–1740.

Clarke, B., Otto, F., Stuart-Smith, R., & Harrington, L. (2022). Extreme weather impacts of climate change: An attribution perspective. *Environmental Research: Climate, 1*(1), 012001.

Dietz, T., Shwom, R. L., & Whitley, C. T. (2020). Climate change and society. *Annual Review of Sociology, 46*, 135–158.

Drash, W. (2017). Yes, climate change made Harvey and Irma worse. *CNN*. https://www.cnn.com/2017/09/15/us/climate-change-hurricanes-harvey-and-irma/index.html

Edwards, T. L., Nowicki, S., Marzeion, B., Hock, R., Goelzer, H., Seroussi, H., Jourdain, N. C., Slater, D. A., Turner, F. E., Smith, C. J., & McKenna, C. M. (2021). Projected land ice contributions to twenty-first-century sea level rise. *Nature, 593*(7857), 74–82.

Eriksen, S., Schipper, E. L. F., Scoville-Simonds, M., Vincent, K., Adam, H. N., Brooks, N., Harding, B., Lenaerts, L., Liverman, D., Mills-Novoa, M., & Mosberg, M. (2021). Adaptation interventions and their effect on vulnerability in developing countries: Help, hindrance or irrelevance? *World Development, 141*, 105383.

Foreign Commonwealth & Development Office (FCDO). (2021). *Climate change adaptation for Caribbean fisheries 2021: Taking action together*. https://www.gov.uk/government/publications/commonwealth-marine-economies-cme-programme-caribbean-region/climate-change-adaptation-for-caribbean-fisheries-2021-taking-action-together

Forster, T., Kentikelenis, A. E., Reinsberg, B., Stubbs, T. H., & King, L. P. (2019). How structural adjustment programs affect inequality: A disaggregated analysis of IMF conditionality, 1980–2014. *Social Science Research, 80*, 83–113.

Fremstad, A., & Paul, M. (2022). Neoliberalism and climate change: How the free-market myth has prevented climate action. *Ecological Economics, 197*, 107353.

Gabriel-Campos, E., Werner-Masters, K., Cordova-Buiza, F., & Paucar-Caceres, A. (2021). Community eco-tourism in rural Peru: Resilience and adaptive capacities to the Covid-19 pandemic and climate change. *Journal of Hospitality and Tourism Management, 48*, 416–427.

Gafar, J. (1978/1979). An analysis of import substitution in a developing economy: The case of Jamaica. *Caribbean Studies, 18*(3/4), 139–156.

Gamso, J. (2017). Trade partnerships and environmental performance in developing countries. *Journal of Environment and Development, 26*(4), 375–399.

Gamso, J. (2023). Aiding animals: Does foreign aid reduce wildlife crime? *Journal of Environment and Development, 32*(1), 34–60.

Girvan, N., Arana Sevilla, M., Ceara Hatton, M., & Rodríguez, E. (1991). The debt problem of small peripheral economies: Case studies from the Caribbean and Central America. *Caribbean Studies, 24*(1/2), 45–115.

Güven, A. B. (2018). Whither the post-Washington Consensus? International financial institutions and development policy before and after the crisis. *Review of International Political Economy, 25*(3), 392–417.

Harvey, D. (2007). *A brief history of neoliberalism*. Oxford University Press.

Hargreaves, T., Hielscher, S., Seyfang, G., & Smith, A. (2013). Grassroots innovations in community energy: The role of intermediaries in niche development. *Global Environmental Change, 23*(5), 868–880.

Hilaire, A. D. L. (2001). Economic stabilization in the Caribbean. *Finance and Development, 38*(1). https://www.imf.org/external/pubs/ft/fandd/2001/03/hilaire.htm

Hossain, M. (2016). Grassroots innovation (GI): A systematic review of two decades of research. *Journal of Cleaner Production, 137*, 973–981.

International Monetary Fund (IMF). (2019). Caribbean tourism in the global marketplace: Trends. https://www.elibrary.imf.org/display/book/9781484315194/ch03.xml

Jerez Columbié, Y. (2022). Adapting to climate change through disaster risk reduction in the Caribbean: Lessons from the Global South in tackling the Sustainable Development Goals. In S. Flood, Y. Jerez Columbié, M. Le Tissier, & B. O'Dwyer (Eds.), *Creating resilient futures: Integrating disaster risk reduction, sustainable development goals and climate change adaptation agendas* (pp. 183–203). Springer.

Jones, S., & Tarp, F. (2016). Does foreign aid harm political institutions? *Journal of Development Economics, 118*, 266–281.

Kemp-Benedict, E., Drakes, C., & Canales, N. (2020). A climate-economy policy model for Barbados. *Economies, 8*(1), 16.

Klack, T. (1997). *Globalization and neoliberalism: The Caribbean context*. Roman & Littlefield Publishers.

Lewis, C. T. (2022). Climate change and the Caribbean: Challenges and vulnerabilities in building resilience to tropical cyclones. *Climate, 10*(11), 178.

Lewis, C. T., & Su, M. C. (2021). Climate change adaptation and sectoral policy coherence in the Caribbean. *Sustainability, 13*(15), 8518.

Liu, Z., Lan, J., Chien, F., Sadiq, M., & Nawaz, M. A. (2022). Role of tourism development in environmental degradation: A step towards emission reduction. *Journal of Environmental Management, 303*, 114078.

Longmore, R., Jaupart, P., & Cazorla, M. R. (2014). *Toward economic diversification in Trinidad and Tobago*. World Bank Policy Research Working Paper No. 6840.
Mackay, E. A., & Spencer, A. (2017). The future of Caribbean tourism: Competition and climate change implications. *Worldwide Hospitality and Tourism Themes, 9*(1), 44–59.
Mahon, R., Petrie, J. A., Trotman, A., Eyzaguirre, J., Burrowes, R., Matthews, L., Van Meerbeeck, C. J., & Charles, A. (2021). Climate services for tourism: Insights from Caribbean small island developing states. *Climate Services, 24*, 100262.
Matallah, S. (2020). Economic diversification in MENA oil exporters: Understanding the role of governance. *Resources Policy, 66*, 101602.
Meditz, S. W., & Hanratty, D. M. (Eds.). (1987). *Caribbean islands: A country study*. GPO for the Library of Congress.
Mohammadi, E., Singh, S. J., McCordic, C., & Pittman, J. (2022). Food security challenges and options in the Caribbean: Insights from a scoping review. *Anthropocene Science, 1*(1), 91–108.
Mohan, P. S. (2023). Financing climate change mitigation and adaptation in Caribbean SIDS. *PLoS Climate, 2*(3), e0000167.
Morgan, E. (2019). CARICOM's economic independence: Role of foreign trade. *CARICOM Today*. https://today.caricom.org/2019/08/07/caricoms-economic-independence-role-of-foreign-trade/9
National Hurricane Center. (n.d.). https://www.nhc.noaa.gov
Pan American Health Organization (PAHO). (2017). *Response to Hurricanes Irma and Maria—2017*. https://www.paho.org/en/health-emergencies/major-emergencies-response/response-hurricanes-irma-and-maria-2017
Parker, L., Bourgoin, C., Martinez-Valle, A., & Läderach, P. (2019). Vulnerability of the agricultural sector to climate change: The development of a pan-tropical Climate Risk Vulnerability Assessment to inform sub-national decision making. *PLoS ONE, 14*(3), e0213641.
Pathak, A., van Beynen, P. E., Akiwumi, F. A., & Lindeman, K. C. (2021). Impacts of climate change on the tourism sector of a Small Island Developing State: A case study for the Bahamas. *Environmental Development, 37*, 100556.
Patnaik, J., & Bhowmick, B. (2020). Promise of inclusive innovation: A re-look into the opportunities at the grassroots. *Journal of Cleaner Production, 259*, 121124.
Peake, S., & Ekins, P. (2017). Exploring the financial and investment implications of the Paris Agreement. *Climate Policy, 17*(7), 832–852.
Peterson, R. R. (2023). Over the Caribbean top: Community well-being and over-tourism in small island tourism economies. *Journal of Community Well Being, 6*, 89–126.
Qian, N. (2015). Making progress on foreign aid. *Annual Review of Economics, 7*, 277–308.
Raghutla, C. (2020). The effect of trade openness on economic growth: Some empirical evidence from emerging market economies. *Journal of Public Affairs, 20*(3), e2081.
Rasmussen, C. (2014). Just 5 questions: Community initiatives against climate change. *NASA Earth Science News Team*. https://climate.nasa.gov/news/1026/just-5-questions-community-initiatives-against-climate-change
Reid, H., Ampomah, G., Olazábal Prera, M. I., Rabbani, G., & Zvigadza, S. (2012). *Southern voices on climate policy choices: Analysis of and lessons learned from civil society advocacy on climate change*. International Institute for Environment and Development.
Rodrik, D. (2006). Goodbye Washington Consensus, hello Washington confusion? A review of the World Bank's "economic growth in the 1990s: Learning from a decade of reform." *Journal of Economic Literature, 44*(4), 973–987.
Roy, D. (2023). How the Caribbean is building climate resilience. *Council on Foreign Relations*. https://www.cfr.org/backgrounder/how-caribbean-building-climate-resilience
Sarvajayakesavalu, S. (2015). Addressing challenges of developing countries in implementing five priorities for sustainable development goals. *Ecosystem Health and Sustainability, 1*(7), 1–4.
Schäfer, M. S., & Painter, J. (2021). Climate journalism in a changing media ecosystem: Assessing the production of climate change-related news around the world. *Wiley Interdisciplinary Reviews: Climate Change, 12*(1), e675.

Schlingmann, A., Graham, S., Benyei, P., Corbera, E., Sanesteban, I. M., Marelle, A., Soleymani-Fard, R., & Reyes-García, V. (2021). Global patterns of adaptation to climate change by Indigenous Peoples and local communities. A systematic review. *Current Opinion in Environmental Sustainability, 51*, 55–64.

Schreuder, W., & Horlings, L. G. (2022). Transforming places together: Transformative community strategies responding to climate change and sustainability challenges. *Climate Action, 1*(1), 1–15.

Shi, L., & Moser, S. (2021). Transformative climate adaptation in the United States: Trends and prospects. *Science, 372*(6549), eabc8054.

Spence, M. (2021). Some thoughts on the Washington Consensus and subsequent global development experience. *Journal of Economic Perspectives, 35*(3), 67–82.

Stiglitz, J. E. (2008). Is there a post-Washington Consensus consensus? In N. Serra & J. E. Stiglitz (Eds.), *The Washington Consensus reconsidered: Towards a new global governance* (pp. 41–56). Oxford University Press.

Thomas, A., Shooya, O., Rokitzki, M., Bertrand, M., & Lissner, T. (2019). Climate change adaptation planning in practice: Insights from the Caribbean. *Regional Environmental Change, 19*, 2013–2025.

Thomas, A., Baptiste, A., Martyr-Koller, R., Pringle, P., & Rhiney, K. (2020). Climate change and small island developing states. *Annual Review of Environment and Resources, 45*, 1–27.

United Nations Environment Program (UNEP). (2023). Adaptation gap report 2023.

U.S. Department of State. (n.d.). Saint Lucia. https://www.state.gov/countries-areas/saint-lucia

Weiler, F., Klöck, C., & Dornan, M. (2018). Vulnerability, good governance, or donor interests? The allocation of aid for climate change adaptation. *World Development, 104*, 65–77.

Williamson, J. (2003). The Washington consensus and beyond. *Economic and Political Weekly, 38*(15), 1475–1481.

Witter, M. (2021). COVID-19: Intensifying the existential threat to the Caribbean. *Agrarian South: Journal of Political Economy, 10*(1), 155–172.

Woodruff, S. C., & Regan, P. (2019). Quality of national adaptation plans and opportunities for improvement. *Mitigation and Adaptation Strategies for Global Change, 24*, 53–71.

Zapata Campos, M. J., Barinaga, E., Dimba Kiaka, R., & Ocampo, J. (2023). Nothing to lose: The rationales of grassroots financial innovations in contexts of extreme scarcity. *Social Enterprise Journal, 19*(2), 193–212.

Chapter 6
A Fuzzy Approach to Assessing Progress Towards the Sustainable Development Goals Using Choquet Integral Aggregation

Mohamed Rami Chouchene, Nihed Mrabet, and Islem Khefacha

Abstract The monitoring of the Sustainable Development Goals (SDGs) poses a significant challenge and strategic opportunity for stakeholders (Sachs et al., in 2018) developed the SDG Index to track progress against a series of goals with multiple targets and indicators. While their index is robust, it overlooks the "balanced" and "integrated" nature of the SDGs. To address these gaps, we propose a new approach that distinguishes between means and end goals using Keeney's theory of value, "Value-Focused Thinking." Additionally, to account for the objective weights of each indicator, we will employ entropy information theory. Using indicators from the official list of SDGs, we will construct indices measuring the achievement of sustainable development goals in a sample of 93 middle- and low-income countries for the period from 2014 to 2018. To consider the interconnected and indivisible nature of the SDGs, we will utilize a fuzzy nonlinear aggregation method known as the Choquet integral. Our model enables the identification of key areas of action for advancing the SDGs. Furthermore, it facilitates the identification of weaknesses in each country, allowing for the prioritization and analysis of relevant policies to address these shortcomings.

Keywords Sustainable development goals · Value-focused thinking · Choquet integral · Multi-criteria analysis

---

M. R. Chouchene (✉) · N. Mrabet
ISFF Sousse, University of Sousse, Sousse, Tunisia
e-mail: ramychouchene@live.fr

N. Mrabet
e-mail: nihed.mrabet@isffs.u-sousse.tn

N. Mrabet · I. Khefacha
LaREMFiQ, IHEC of Sousse, University of Sousse, Sousse, Tunisia
e-mail: Islem.Khefacha@fsegma.u-monastir.tn

I. Khefacha
FSEG Mahdia, University of Monastir, Monastir, Tunisia

M. S. Ben Ali and E. Lechman (eds.), *Sustainable Economic Development*,
https://doi.org/10.1007/978-981-97-3767-3_6

JEL Q01 · D46 · D81

6.1 Introduction

Scholars, policymakers, and various stakeholders have actively engaged in intense discussions concerning the metrics used to measure sustainability. The introduction of the 2030 Agenda, along with its Sustainable Development Goals (SDGs), has significantly elevated the discourse on this subject. The primary objective of the 2030 Agenda is to promote a more prosperous and enduring future for all. The interconnectedness of its three main components, economic, social, and environmental, distinguishes the sustainable development agenda (Bali Swain & Ranganathan, 2021; Dang & Serajuddin, 2020; Moyer & Hedden, 2020). This framework has paved the way for the establishment of the Sustainable Development Goals, embodying integrated and indivisible objectives with global and universally applicable significance (Biggeri et al., 2019; Dawes, 2022; Heck et al., 2018).

The UN's 17 Sustainable Development Goals (SDGs), comprising 169 objectives and 232 indicators, form the foundation of the agreement. The SDGs aspire to establish a harmonious equilibrium between social inclusivity, environmental sustainability, and economic prosperity. They represent a set of interconnected global objectives, necessitating an understanding of how the goals influence each other and consistency in their formulation, implementation, and monitoring (Biggeri et al., 2019; Coenen et al., 2021; Liu, 2020; Moyer & Hedden, 2020).

The approach to matching and monitoring the achievements of Sustainable Development Goals (SDGs) should illustrate the evident interdependence and mutual reinforcement among objectives and targets at various levels of implementation, both in theory and practice (Bennich et al., 2023; Kostetckaia & Hametner, 2022; Sam, 2021; Yu & Huarng, 2024). A thorough comprehension of potential advantages and disadvantages, as well as the interrelationships among different objectives, is crucial. It is imperative to assess how the attainment or non-attainment of a specific goal may yield both positive and negative effects on other goals and to what extent (Guo et al., 2023; Issa et al., 2024; Kuc-Czarnecka et al. 2023; Renaud and al., 2022).

The way that the Sustainable Development Goals (SDGs) are matched and tracked should show how clearly objectives and targets are interdependent and reinforce each other at different levels of implementation, both in theory and in practice (Bennich et al., 2023; Kostetckaia & Hametner, 2022; Sam, 2021; Yu & Huarng, 2024). A thorough understanding of potential advantages and disadvantages, as well as the interrelationships among different objectives, is crucial. It is imperative to assess how the attainment or non-attainment of a specific goal may yield both positive and negative effects on other goals and to what extent (Issa et al., 2024; Guo et al., 2023; Kuc et al., 2023; Renaud et al., 2022).

Certain critics argue for the reduction of the Sustainable Development Goals (SDGs), citing perceived impracticality, idealism, and a lack of clarity (Dang & Serajuddin, 2020; Dawes, 2022; Issa et al., 2024). However, it is crucial to recognize that assessing the SDGs should extend beyond their conceptual or technical precision. The SDGs are designed as a strategic framework to promote collaboration and collective action among diverse stakeholders and decision-makers, all with the aim of realizing a more promising future.

Additionally, it is inaccurate to assume that the interdependence of the 17 objectives and 169 targets is contingent on the context. Addressing this matter, Kanbur and Stiglitz (2016) proposed that the SDGs provide a rational foundation for selecting and improving a set of measures. They argue that there are valid reasons for limiting the number of indicators that can be displayed on dashboards.

On the flip side, there has been a notable proliferation in the number of goals, targets, and indicators, leading to criticism about the ability to oversee, quantify, and manage such an extensive, intricate, and interconnected array of advancements (Issa et al., 2024). One approach to establishing a monitoring framework for the Sustainable Development Goals (SDGs) involves concentrating on specific targets at local and sectoral levels (Chen et al., 2023; Fang et al., 2023; Kynčlová et al., 2020). Another strategy includes advocating for the use of a subset reduced to subobjectives. However, these options prove ineffective for monitoring "global" aspects of development, such as climate issues or deforestation, and for making international comparisons. Recent studies have initiated the development of comprehensive measures utilizing SDG indicators for specific sectors. For instance, Brelsford et al. (2017) focused on the urban environment; Giupponi and Gain (2016) examined water, energy, and food; and Nhemachena et al. (2018) explored agriculture.

Monitoring the progress of the Sustainable Development Goals (SDGs) presents a substantial challenge and strategic advantage for all stakeholders and beneficiaries participating in the 2030 Agenda, regardless of their level of engagement (Issa et al., 2024; Nerland et al., 2022; Sun et al., 2018). The Sustainable Development Objectives Index, crafted by Jeffrey Sachs and colleagues, aims to track advancements across a spectrum of goals, incorporating diverse targets and indicators while monitoring overall development. While effective for country comparisons, the index falls short in considering the "balanced" and "integrated" aspects of the SDGs. Moreover, it grapples with well-known issues associated with utilizing an arithmetic average, assuming a linear relationship between dimensions, and overlooking potential trade-offs and synergies. Additionally, the SDGs provide a harmonious blend of desired outcomes and approaches to achieving them. Consequently, some researchers (Haq, 1995; Streeten, 1994; Stern et al., 2022), along with Stiglitz et al. (2019), underscore the importance of distinguishing between the developmental objectives and the strategies employed to achieve these objectives.

To address these gaps, this article conducts an in-depth analysis of the numerous challenges associated with the identification and measurement of the Sustainable Development Goals (SDGs), with a primary focus on quantitative methods. The analysis encompasses global data, with a specific emphasis on statistics from economically disadvantaged and middle-income countries. We delve into fundamental issues often assumed to be resolved, revealing potential variations upon closer examination.

Moreover, we seek to address questions regarding the feasibility of quantifying progress towards achieving the SDGs. Are there alternative methods to effectively monitor this progress? Can we track a set of core indicators instead of relying on a comprehensive index? How does the absence of statistical data for certain indicators, a widely acknowledged reality, impact the evaluation of a country's level of achievement? Moreover, we explore the distinction between the inputs and outputs of the SDGs and consider the most appropriate method of combining data that accounts for this phenomenon while weighing the potential advantages and disadvantages inherent in the relationships between goals and targets.

We propose an innovative approach that distinguishes between means and ultimate goals by applying Keeney's theory of values. This technique aims to address the aforementioned problems (Keeney, 1996). The Choquet integral is the specific fuzzy nonlinear aggregation approach that we shall employ in the upcoming stages (Pinar, 2021). Furthermore, we will utilize the entropy information theory to account for the objective weights linked to each indicator. Therefore, based on the current information available, our work is focused on developing a new approach to restructuring the SDGs by categorizing them into goals of ends and goals of means, using Keeney's theory of value. Consequently, it suggests a novel measure of achievement for the Sustainable Development Goals by considering possible trade-offs.

The remaining sections are organized as follows. Measuring sustainability and SDG interlinkages are discussed in Sect. 6.2. Then, we describe the data and methodology, present and discuss our results, and conclude with a synthesis of our findings, areas for future research, and suggestions for further study.

6.2 Literature Review

Several authors have dedicated their efforts to the creation and implementation of model frameworks designed to address intricate sustainability challenges. Notably, Kılkış (2016) devised a composite index specifically tailored for evaluating energy, water, and environmental systems in Southeast European cities. Addressing the sustainability of urban living, Valcárcel-Aguiar et al. (2018) proposed a composite index as well. In the context of comprehending the diverse developmental phases observed in African nations, Bartniczak and Raszkowski (2018) utilized an indicator-based approach. Meanwhile, Mononen et al. (2016) developed a plan for assessing ecosystem services, focusing on the Finnish context. Examining the interconnected dynamics of environmental pollution, economic growth, and public health in China, Lu et al. (2017) conducted a comprehensive study. Additionally, Pinar et al. (2014) contributed to the field by constructing a sustainable development index, incorporating 19 distinct indicators in their methodology.

Moreover, traditional indicators continue to be widely employed for quantifying specific aspects of sustainable development, despite their limitation in fully integrating the interdisciplinary perspectives inherent in the various dimensions of sustainability. Metrics such as the Human Development Index (Cohen, 2004), the GINI coefficient of income inequality (Gini, 1921), and the Social Progress Index

(Stern et al., 2022) are examples of such conventional measures utilized in assessing development.

Sachs et al. (2018) developed a specialized indicator designed exclusively for evaluating the Sustainable Development Goals (SDGs). This index utilizes a system in which the value is determined by calculating the average of all indicators for each objective. As a result, each indicator is assigned the same level of importance, guaranteeing consistency across all countries. Put simply, governments are prohibited from giving more weight to some indicators than others, which hinders the development of priorities (Weber & Weber, 2020; Breuer et al., 2023; Saiu et al., 2022; Wong & Heijden, 2019). The measure is used to assess a country's performance relative to all other countries involved in the research.

In addition, the OECD study provides a different method for evaluating the effectiveness of the Sustainable Development Goals (OECD, 2017). This approach entails the computation and determination of the standardized distance from a desired outcome for each metric, providing a strategy that enhances the assessment procedure. Another method to evaluate the effectiveness of the Sustainable Development Goals is to utilize the monitoring report generated by Eurostat on SDG progress (Eurostat, 2021). Under the premise of ongoing trends, a projection is generated for each country for the year 2030 in this approach. The start and end values of each indicator are used as observations to calculate growth rates. This is done by extrapolating linearly from the initial values to the final values in 2030. The values for 2030. derived from the advancement of indicators and employing uniform weighting, are thereafter incorporated into the targets. This approach offers two notable advantages. Firstly, it enables the incorporation of several additional indicators without requiring explicit objectives. Furthermore, utilizing Eurostat data allows for an assessment of the effectiveness of achieving the Sustainable Development Goals. According to a United Nations report tracking progress towards the 2030 Sustainable Development Goals (UN, 2018), since the year 2000. there have been notable reductions in rates of extreme poverty, maternal mortality, and infant mortality.

Simultaneously, there has been an increase in the percentage of the global population with access to electricity. However, the report also highlights concerning trends, such as the rise in developing countries' per capita material footprint and a decrease in the global share of sustainably exploited stocks. According to Barbier and Burgess (2019), the fight against poverty may have progressed at the expense of other environmental and social sustainable development goals from 2000 to 2015. The body of literature addressing the utilization of systems thinking and analysis for prioritizing national goals and assessing interdependencies among Sustainable Development Goals (SDGs) is expanding. Barbier and Burgess (2019) have developed a method to measure the interrelationships between objectives using a welfare function. Additionally, Obersteiner et al. (2016) have identified that policies related to Sustainable Consumption and Production (SDG 12) play a pivotal role in minimizing trade-offs between environmental measures and food costs. Pradhan et al. (2017) systematically examined data from 227 countries to assess correlations between sustainable development indicators. They categorize a statistically significant positive correlation between two sustainability indicators as a synergy, indicating a mutually beneficial relationship. Conversely, a trade-off is defined as a statistically significant negative correlation between two pairs of indicators, suggesting a situation where improvements in one aspect may come at the expense of another. Ivanov and Peleah (2018) have innovatively developed a conceptual framework aimed at assessing progress towards sustainable development. Drawing inspiration from the established Human Development Index (HDI), their framework incorporates considerations for trade-offs and complementarities. Expanding on the HDI, traditionally focused on health, education, and living standards, the authors introduce a fourth environmental dimension. This broader HDI index captures both the state and process of sustainable human development while considering crucial environmental factors. The SDGs have faced criticism for their perceived lack of specificity and coherence, with some parts appearing to lack compatibility. The International Council for Science (ICSU) has expressed concern about SDG incompatibility, particularly in the area of socioeconomic development versus environmental sustainability. As exemplified by the works of Spaiser et al. (2016), Moyer and Hedden (2020), and Biggeri et al. (2019), numerous studies have attempted to quantify and analyze contradictions, conflicts, and interconnections within the SDGs. For instance, Spaiser et al. (2016) employ dynamical systems models to delve into the origins of these discrepancies, revealing that the pursuit of economic growth and consumption as methods for development contributes to the observed inconsistencies. The results of these studies show how important it is to think about how Sustainable Development Goals (SDGs) affect each other when making plans and decisions (Cling & Delecourt, 2022; Dawes et al., 2022; Guo et al., 2023; Huang, 2023; Horn & Grugel, 2018; Tosun & Leininger, 2017; Kroll et al., 2019; Liu, 2020; Osman et al., 2022; Pingali & Plavšić, 2022; Schmidt et al., 2017; Weitz et al., 2017; Zhu et al., 2022). Separating the measuring and monitoring of Sustainable Development Goals (SDGs) can affect the successful implementation of government objectives. According to Nilsson et al. (2018), different government agencies are responsible for supervising distinct sectors of public life.

Policymakers need relevant information and resources to identify important trade-offs and synergies and monitor their progress or decline from the overall Sustainable Development Goals (SDG) objective. Hence, by identifying the connections between various Sustainable Development Goal (SDG) goals, it becomes possible to create a decision framework that may prioritize objectives and facilitate strategic planning (Asadikia et al., 2021; Guerrero et al., 2023; Allen et al., 2021). Le Blanc (2015) examines the relationship between each Sustainable Development Goal (SDG) and its corresponding objectives. The study focuses on creating an interlinkage matrix that illustrates how each target is connected to other targets and how it supports the fundamental objective of the SDG. His analysis demonstrates that the interconnections between SDG12 (responsible consumption and production) and other objectives are notably strong, but SDG14 (life below water) has just two such interconnections. There exist compelling intellectual and practical justifications for systematically and interactively addressing the complexity of development.

According to Haq (1995), Hicks and Streeten (1979), Stern et al. (2022), Streeten (1994), and Torlockyj (1975), various publications have emphasized the importance of differentiating between the inputs and outputs of development processes. Additionally, academics like, Biggeri and Mauro (2018), Clark (2005), Sen (1999) and have emphasized the intricate relationships and interactions between various capacities and freedoms. Nevertheless, even the most basic dashboards may not be the most effective tools for the task at hand. Researchers like Clark (2016), Schmidt-Traub et al. (2017) and Singh and Chudasama (2021) contend that a single aggregate measure, such as the SDG Index, provides a straightforward and clear summary of overall performance. This simplicity is valuable, considering the challenge of differentiating results from analyses involving numerous targets and indicators that often overlap and move in different directions. A single aggregate measure, such as the SDG Index, proves beneficial by providing a cohesive overview. In other words, as Costanza et al. (2016) state, while having a well-equipped dashboard is vital, it is equally crucial to understand the direction and progress towards the objective. They argue that a single aggregate measure could stimulate public discourse more effectively than presenting numerous individual results that might lead to selective interpretation.

6.3 Research Methodology

The objective of this section is to present a model for evaluating the attainment of the Sustainable Development Goals, considering the challenges and gaps identified in the preceding section. To commence, we will differentiate between the SDGs as end goals and means goals, employing the theory of value-focused thinking to delineate between policies and strategies for implementation and the ultimate goals of development. Subsequently, we will consolidate all the indicators of the Sustainable Development Goals into a unified measure, considering their trade-offs and interconnections. To achieve this, we will employ the Choquet integral, a fuzzy nonlinear aggregation method.

6.3.1 New Structure of the Sustainable Development Goals

As discussed in the previous section, the 2030 Agenda comprises 17 goals and 169 targets designed for proactive measures. The primary aim of this agenda is to realize sustainable development by the year 2030. Achieving sustainable development poses challenges due to its inherent interconnectedness, dependence on various factors, complexity, and the presence of conflicting goals. Using Keeney's (1996) Value-Focused Thinking (VFT) theory, we suggest rearranging the Sustainable Development Goals to address these issues.

The Value-Focused Thinking (VFT) theory revolves around the concept of how values can be employed to enhance decision-making processes. According to Keeney (1996), values lie at the core of all our actions and should serve as the guiding force behind decision-making. Scholars like Kunz et al. (2016) and Shukla et al. (2018) propose that VFT offers a systematic approach to structuring intricate decisions and subsequent analyses, providing valuable insights for effective decision-making.

Françozo and Belderrain (2022) have demonstrated that embracing values enhances the decision-making process, which comprises three integral components: the decision's context, an object, and a preference direction. By incorporating values into this process, decision-makers gain a more comprehensive and structured framework for tackling complex decisions, leading to more informed and aligned choices. VFT, therefore, emerges as a valuable tool for navigating decision complexity and ensuring that values play a central role in the decision-making journey.

Initiating an analysis of Value-Focused Thinking requires unveiling the decision-making structure and encompassing the subject's context and primary goals. Within a specific decision scenario, the problem context defines the array of suitable alternatives for consideration. It is through the exploration of these alternatives that the core values intrinsic to the decision are uncovered. In essence, the decision-making structure of Value-Focused Thinking hinges on understanding the contextual factors, establishing overarching goals, and discerning key values by examining a range of appropriate alternatives.

The ensemble of 17 Sustainable Development Goals (SDGs) and 169 associated targets forms the foundation for our decision-making processes. Currently, these goals lack a structured framework. The subsequent step involves delineating between end goals, which are foundational, and means goals. As asserted by Keeney (1996), the fundamental goal serves as the crux of why the decision-making situation holds significance. In parallel, he characterizes means goals as an expression of interest in the decision context, driven by its impact on the likelihood of achieving a distinct, more pivotal goal. In essence, means goals function as the requisite tools to realize primary objectives. Keeney's approach to distinguishing between end goals and means goals revolves around the fundamental question, "Why is this goal important?" This inquiry serves as a guiding principle for elucidating the inherent importance of each goal within the decision-making context.

In the context of the Sustainable Development Goals, decision-makers are interested in how well the overall goal is being met. As we mentioned in the previous section, the aim of sustainable development is to create consistent, inclusive, and long-lasting economic growth, a world without any kind of poverty, social cohesion, and the preservation of species, natural resources, and energy. To do this, we will assess the effectiveness of development in three ways: economic, social, and environmental.

- Economic development (ECOD): Growth that is long-lasting, inclusive, and sustainable creates wealth over time. We suggest four sub-dimensions to quantify it: economic growth, macroeconomic stability, sustainable consumption and production, and inclusion.

- Social Development (SOCD): The social dimension of sustainable development looks at how many people are also included and how society works as a whole. For this, we looked at how the Social Progress Index is defined, which is described as follows: "Social progress is a society's capacity to satisfy the fundamental human requirements of its citizens (basic human needs), to develop the building blocks that allow people and communities to enhance and sustain their quality of life (foundations of well-being), and to provide the environment for all individuals to reach their full potential (opportunities)". As indicated in Table 6.1, the dimension of social performance is determined by three sub-dimensions: opportunities, the foundations of well-being, and basic human needs.
- Environmental Development (ENVD): The third component of sustainable development seeks to quantify the degree of preservation of biodiversity (fauna and flora) and natural and energy resources, as well as air pollution. It is measured along two sub-dimensions: air quality (ENV1) and natural resource endowment (ENV2) (see Table 6.1).

On the other hand, good governance, sound policies (economic, social, and environmental), and reinforcing human and financial capacity are the ways that sustainable development can be put into action to reach its goal. There are three sub-dimensions that make up the means of implementation dimension. Three parts look at the policy sub-dimension (POL) of sustainable development (economic, social, and environmental) (SOCP, ENVP, and ECOP). One measure is the quality of governance and institutions (GOV). The third dimension is about strengthening financial and human capacity (FHC) (Table 6.2).

To distinguish between end goals and means goals and assess our proximity to achieving the Sustainable Development Goals (SDGs), we will employ the UN's official list of SDG indicators (United Nations, 2019). The SDG Achievement Index variables align, whenever feasible, with the officially approved SDG indicators endorsed by the United Nations Statistical Commission. To ensure the indicators' quality meets the required standards, we apply four selection criteria:

- Applicable globally and adaptable across diverse countries, these indicators serve the purpose of tracking the Sustainable Development Goals (SDGs) universally. Their uniformity ensures consistency across continents, facilitating straightforward comparisons of countries' progress. Notably, these indicators enable the establishment of numerical performance thresholds, offering a clear gauge of the extent to which sustainable development goals are being achieved.
  - Statistical robustness: The chosen indicators are robust and dependable measures.
  - Punctuality: The selected indicators are current and are released within a reasonable timeframe.
  - Data quality: Data sets, sourced from official national or international entities (such as national statistical offices or international organizations) or other reputable sources like peer-reviewed publications, represent the optimal means to measure specific issues reliably.

Table 6.1 Evaluation system of end goals of sustainable development performance

Dimensions	Sub index	Basic indicators	Proxy variables	SDG goals
Economic performance (ECOD)	Economic growth (ECO1)	Economic growth	Annual growth rate of real GDP per capita (%)	G0811
	Macroeconomic stability (ECO2)	Debt services	Debt service as a proportion of exports of goods and services (%)	G1741
		Current account balance	Current account balance	G17132
		Economic growth volatility	Volatility of the annual growth rate of real GDP	G17133
		Inflation	Inflation, consumer prices (annual %)	G171313
	Sustainable consumption and production (ECO3)	Energy intensity	Level of energy intensity of primary energy (megajoules per GDP at constant 2011 purchasing power parity)	G0731
		Material consumption	Domestic material consumption per unit of GDP (kilograms per constant 2010 US dollar)	G0842
	Economic inclusion (ECO4)	Poverty	Proportion of population living below the international poverty line ($1.9 per day)	G0111
		Income Inequality	The Gini index	G1041
		Unemployment	Unemployment rate (%)	G0851
Social performance (SOCD)	Opportunities (SOC1)	Capability building	People using the Internet (% of population)	G1781
			Postgraduate enrolment rate (% gross)	G0431
			Proportion of adults (15+) with an account at a financial institution or mobile money service provider (% of adults aged 15+)	G08103
		Women empowerment	Percentage of contributing family workers, women (% of female employment)	G0542
			Proportion of seats held by women in national parliaments (%)	G0551

(continued)

Table 6.1 (continued)

Dimensions	Sub index	Basic indicators	Proxy variables	SDG goals
			Gender parity index of enrolment rate in primary education (gross)	G0451
			Gender parity index of enrolment rate in secondary education (gross)	G0452
			Gender parity index of higher education enrolment rate (gross)	G0453
	Foundations of Well-being (SOC2)	Health and well-being	the mortality rate attributed to cardiovascular diseases, cancer, diabetes, or chronic respiratory diseases	G0341
			Suicide mortality rate per 100,000 inhabitants	G0342
			Alcohol consumption per capita (15 years and over) during a calendar year (litres of pure alcohol)	G0351
			The mortality rate attributed to unintentional poisonings (deaths per 100,000 inhabitants)	G0391
		Access to basic knowledge	School enrolment rate, pre-primary (% gross)	G0421
			Literacy rate (% among people aged 15 and over)	G0461
		Education quality	Out-of-school adolescent rate (% lower secondary age)	G0411
			Rate of out-of-school children (% of primary school age)	G0412
	Basic human needs (SOC3)	Shelter	The proportion of the population with access to electricity	G0711
			Proportion of population primarily dependent on clean fuels and technology	G0712
			Proportion of urban population living in slums	G1111
		Water and sanitation	Proportion of the population using basic water services	G0142

(continued)

Table 6.1 (continued)

Dimensions	Sub index	Basic indicators	Proxy variables	SDG goals
			Proportion of the population using basic sanitation services	G0141
			Proportion of the population practicing open defecation	G0621
		Nutrition and basic medical care	Prevalence of undernourishment (%)	G0211
			Maternal mortality rate	G0311
			Number of new HIV infections per 1000 uninfected inhabitants	G0331
			Incidence of tuberculosis per 100,000 inhabitants	G0332
			Incidence of malaria per 1000 people at risk	G0333
			Under-five mortality rate per 1000 live births	G0321
			Neonatal mortality rate per 1000 live births	G0322
			Proportion of children with moderate or severe stunting	G0221
			Proportion of moderately or severely overweight children	G0222
			Proportion of moderately or severely wasted children	G0223
Environmental performance (ENVD)	Air quality (ENV1)	CO_2 emission	Carbon dioxide emissions per unit of GDP (kilograms of CO_2 per constant 2010 US dollars)	G0941
		Air pollution	Average annual PM2.5 air pollution exposure (micrograms per cubic meter)	G1161
	Natural resource endowment (ENV2)	Freshwater withdrawals	Freshwater withdrawals as a proportion of available freshwater resources	G0642
		Renewable energy	Share of renewable energy in total final energy consumption	G0721
		Preservation of biodiversity	Net forest area exchange rate	G1521

(continued)

Table 6.1 (continued)

Dimensions	Sub index	Basic indicators	Proxy variables	SDG goals
			The proportion of degraded land in the total area	G1531
			Red list index	G1551

To explore a strong dataset and deal with outliers, we considered filling in the gaps in the missing parts of the data. Indeed, we calculated the five-year average between 2014 and 2018. In addition, the unconditional average imputation was used to determine missing values in terms of the World Bank income group classification for the data set. The formula below was used to obtain a complete data set:

$$\overline{x}_{j,i} = \frac{1}{n_{j,i}} \sum x_{j,i}$$

The formula above indicates the average values attributed to indicator j for countries in class i. Moreover, it represents the random variable of the indicator j for the countries of classes i. Thus, the number of non-missing values in index j for countries in class i.

Since each metric is not only measured using different methods, but different levels correspond to different sustainability outcomes, we first normalize them using a modified mix/max technique developed for all metrics before aggregating them into a single composite index. The following normalization process is applied to the positive and negative direction criteria, respectively, to obtain the normalized assessment score for each indicator.

$$a_{ij} = \frac{x_{ij} - \min(x_i)}{\delta_i - \min(x_i)}$$

or

$$a_{ij} = \frac{\max(x_i) - x_{ij}}{\max(x_i) - \delta_i}$$

Knowing that x is the observation of variable i and country j is the target of SDG i: if the indicator is in the positive sense, δ is considered the maximum value, whereas if the indicator is in the negative sense, δ is taken as the minimum value.

The a_{ij} value is stored between [0, 1]. Any value <0 is considered 0. So, any value >1 is considered one. In addition, these normalized values have become easy to interpret and compare across all indicators. A country that scores 0.5 for a variable is halfway to achieving the optimal value; a country with a score of 0.75 has covered three-quarters of the distance from worst to best.

Table 6.2 Evaluation system of mean goals of sustainable development performance

Dimensions	Sub-dimensions	Sub index	Proxy variables
Mean goals (MEAN)	Policies (POL)	Economic policies (ECOP)	Merchandise trade (% of GDP)
			Manufacturing value added as a percentage of GDP (%)
			Proportion of the medium and high technology industries' added value in the total added value (%)
			Annual growth rate of real GDP per worker (%)
			Gross capital formation (% of GDP)
		Social Policies (SOCP)	Proportion of target population with access to 3 doses of diphtheria, tetanus, and pertussis (DTP3) (%)
			Proportion of target population with access to a second dose of measles virus-containing vaccine (MCV2) (%)
			Proportion of the target population with access to the third pneumococcal conjugate vaccine (PCV3) (%)
			Number of ATMs per 100,000 adults
			Number of commercial bank branches per 100,000 adults
		Environmental policies (ENVP)	Degree of implementation of integrated water resources management (%)
			Proportion of sites importance for terrestrial and freshwater biological diversity covered by protected areas
			Total pre-tax fossil fuel subsidies (consumption and production) as a share of GDP (%)
	Governance (GOV)	The ability of government to effectively formulate and implement sound policies. (OUT)	Government effectiveness
			Regulatory quality

(continued)

Table 6.2 (continued)

Dimensions	Sub-dimensions	Sub index	Proxy variables
		The respect of citizens and the state for the institutions that govern economic and social interactions between them (CXT)	Rule of Law
			Fight against corruption
		The process by which governments are selected, controlled, and replaced (IMP)	Political stability and absence of violence/terrorism
			Voice and Accountability
	Financial and Human Capacity (FHC)	Human Capacity (HC)	Density of health workers, nurses (per 10,000 inhabitants)
			Density of health workers, physicists (per 10,000 inhabitants)
			Primary pupil/teacher ratio
		Financial Capacity (FC)	Foreign direct investment, net inflows (% of GDP)
			Volume of remittances (in US dollars) as a proportion of total GDP (%)
			Net development aid received (% of GNI)
			Total government revenue (budgetary central government) as a share of GDP (%)

6.3.2 Choquet Integral as an Aggregation Function of a Multi-criteria Problem

Assessing the level of accomplishment of the Sustainable Development Goals (SDGs) is a challenging task due to the complexity and interconnected structure of the problem, which consists of several criteria. The interconnectedness of these factors presents a barrier in the process of assessment. Several approaches have been suggested in the academic literature to tackle the intricacies of multi-attribution difficulties. Each of these strategies is based on unique mathematical principles, resulting in different outcomes. In this part, we will quickly examine the various methodologies for categorizing items in order to assess them. Some of these techniques are inadequate for addressing sustainability challenges. In this regard, we emphasize the deficiencies of certain categorization methods in effectively addressing the complexities of sustainability assessment. Therefore, we strongly support the use of the Choquet integral as a preferable approach for combining sustainability indices. The selection is based on its capacity to handle the complex connections and interactions that are inherent in sustainability criteria, making it a more effective option.

Thomas and Deakin (2017) categorized multi-attribute aggregation approaches into three distinct groups: multi-attribute value theory (MAVT), over-ranking approaches, and interactive approaches. Among these, MAVT methods are widely utilized when dealing with problems featuring multiple attributes. These methods employ aggregation algorithms to assign scores to each option under consideration.

In MAVT-based approaches, the criteria are assumed to be mutually preferentially independent, and an additive utility function is frequently employed. Usually, to find out how useful or satisfying each option is overall, you add up how well each alternative meets each criterion and how important or weighted the criteria are in relation to the decision problem's goals. Despite the simplicity and clarity of MAVT methods, the assumption of criteria independence may not be the most effective approach for resolving multi-attribute problems, especially when the interaction between criteria plays a crucial role in decision-making.

The decision framework centered around the Choquet integral is a frequently employed method for addressing interactions among criteria in scenarios involving multiple criteria. As a non-additive integral, it functions as an aggregation operator to determine the overall satisfaction of each alternative (Choquet, 1954). In contrast to traditional methods rooted in Multi-attribute Value Theory (MAVT), Choquet's integral methods have demonstrated superior effectiveness in handling multi-criteria problems, particularly when criteria exhibit interdependence. It is crucial to highlight that within this framework; the weighting of criteria and interaction coefficients typically relies on the subjective opinions or preferences of decision-makers.

However, when confronted with intricate issues in multi-criteria analyses, such as the assessment of the Sustainable Development Goals (SDGs), where numerous criteria are interconnected and the interactions are nuanced, defining the subjective preferences of decision-makers regarding criteria and weights can be challenging and inconsistent. This presents a research gap in multi-criteria problems, particularly concerning the weighting of interdependent criteria, raising the question of how to objectively determine the weights of interdependent criteria solely based on data without relying on the subjective opinions of decision-makers.

To address this challenge, we propose employing an optimization model method to objectively and impartially determine interaction coefficients and weights for criteria across various levels. This method leverages the theories of fuzzy measures, Shapley's value, and Marichal's entropy, integrating objective data to ensure a robust and unbiased approach (Liu et al., 2015). The utilization of objective weight data serves to safeguard against the influence of inconsistent or uncertain subjective judgments, thereby maintaining the reliability and neutrality of the assessment. This approach is particularly effective for solving multi-criteria decision problems where the reliable and subjective determination of interactions and weights for interdependent criteria proves challenging.

The evaluation process includes the following key steps:

- **Step 1**: Get the a_{ij} values ranged between [0, 1] using: $a_{ij} = \frac{x_{ij} - \min(x_i)}{\delta_i - \min(x_i)}$
- **Step 2**: Discretization of a_{ij} into 5 classes (0, 0.25, 0.5, 0.75, 1) corresponding to the following intervals: [0, 1/5], [1/5, 2/5], [2/5, 3/5], [3/5, 4/5], [4/5, 5/5].
- **Step 3**: Get the contribution of each criterion ($\mu(S)$) using: $\mu(S) := \frac{h\{X_{i_1}, \ldots, X_{i_S}\}}{h(\aleph)}$ si $S = \{A_{i_1}, \ldots, A_{i_S}\}$
- **Step 4**: Obtain the objective weight of each criterion using the maximization of Marichal entropy:

$$H_M(\mu) = -\sum_{i=1}^{m} \sum_{T \subseteq M \backslash \{A_1\}} \frac{(m - |T| - 1)!|T|!}{m!}$$
$$* \left[\mu(T \cup \{A_i\}) - \mu(T)\right] ln\left[\mu(T \cup \{A_i\}) - \mu(T)\right]$$

- **Step 5**: Obtain the evaluation score aggregated by the Choquet integral using the equation:

$$C_\mu(a) := \sum_{i=1}^{m} a_{(i)} \left[\mu(B_{(i)}) - \mu(B_{(i+1)}) \right].$$

6.4 Results and Interpretation

In this section, we will present the results of the aggregation of the SDG achievement index in order to give recommendations for minimizing the cost of implementing the achievement of these goals.

The Choquet integral methodology involves the identification of the set of capacities, or weights, given to subsets of dimensions. In this subsection, we first propose the weights for each dimension of the SDG achievement index, in particular, the end goals (sustainable development) and the means goals (means of implementation). In addition to the dimensions of the SDG achievement index, we present level 3 and 4 weights. As described in previous sections, when a node has 3 criteria, 8 combinations of criteria are possible.

Consequently, it is evident that the sum of the weights of the criteria within each dimension does not equal one. As indicated in the Table 6.3, there are interacting links between the dimensions, the criteria under each dimension, and the sub-criteria under each criterion. We also observe that all the interaction indices are negative, which is due to the submodular entropy capacity of the Choquet integral. And as a result, we can say that all dimensions, criteria, and variable sub-dimensions are competitive.

Table 6.3 The weights of the dimensions and sub-dimensions of the SDGAI

End goals		End goals		End goals	
Sustainable development	Weight	Means	Weight	SDG	Weight
{}	0.000	{}	0.000	{}	0.000
{ECOD}	0.345	{POL}	0.302	{SD}	0.429
{SOCD}	0.426	{GOV}	0.374	{MEAN}	0.657
{ENVD}	0.364	{FHC}	0.438	{SD, MEAN}	1.000
{ECOD, SOCD}	0.728	{POL, GOV}	0.637		
{ECOD, ENVD}	0.688	{POL, FHC}	0.701		
{SOCD, ENVD}	0.733	{GOV, FHC}	0.770		
{ECOD, SOCD, ENVD}	1.000	{POL, GOV, FHC}	1.000		
Economic development		Policies			
{}	0.000	{}	0.000		
{ECO1}	0.393	{SOCP}	0.415		
{ECO2}	0.303	{ENVP}	0.342		
{ECO3}	0.321	{ECOP}	0.370		
{ECO4}	0.227	{SOCP, ENVP}	0.727		
{ECO1, ECO2}	0.668	{SOCP, ECOP}	0.751		
{ECO1, ECO3}	0.690	{ENVP, ECOP}	0.690		
{ECO1, ECO4}	0.602	{SOCP, ENVP, ECOP}	1.000		
{ECO2, ECO3}	0.599	Governance			
{ECO2, ECO4}	0.515	{}	0.000		
{ECO3, ECO4}	0.527	{OUT}	0.431		
{ECO1, ECO2, ECO3}	0.918	{CXT}	0.357		
{ECO1, ECO2, ECO4}	0.848	{INP}	0.439		
{ECO1, ECO3, ECO4}	0.851	{OUT, CXT}	0.657		
{ECO2, ECO3, ECO4}	0.771	{OUT, INP}	0.817		
{ECO1, ECO2, ECO3, ECO4}	1.000	{CXT, INP}	0.728		
Social development		{OUT, CXT, INP}	1,000		
{}	0.000	Building capacity			
{SOC1}	0.407	{}	0.000		
{SOC2}	0.400	{HC}	0.549		
{SOC3}	0.399	{FC}	0.480		

(continued)

Table 6.3 (continued)

End goals		End goals		End goals	
{SOC1, SOC2}	0.740	{HC, FC}	1.000		
{SOC1, SOC3}	0.730				
{SOC2, SOC3}	0.708				
{SOC1, SOC2, SOC3}	1.000				
Environmental development					
{}	0.000				
{ENV1}	0.506				
{ENV2}	0.535				
{ENV1, ENV2}	1.000				

For the dimension of the achievement of the SDGs, we observe that the dimension of the means of implementation is more important than the dimension of sustainable development. Consequently, their better performance will allow acceleration towards the achievement of the Sustainable Development Goals.

In the same way, we observe that the human and financial capacity sub-dimensions are the most important means of implementation (0.438), followed by governance (0.374), and policies are the least important dimensions. So, in order to achieve the SDGs, great attention must be given to building capacity, whether human or financial (0.549; 0.480).

Regarding the end goals, Table 6.3 shows that the social pillar is more important than the environmental and economic pillars (0.426; 0.364; 0.345 successively). At the level of the sub-dimension of sustainable development, we observe that the adoption of a better opportunity and better search for well-being will allow for better social development and better sustainable development. This confirms Sen's capability theory.

For economic development, high economic growth and better allocation of resources lead to better economic development since their weights are successively 0.393 and 0.321.

For the last dimension of environmental development, air quality and natural resource endowment are almost equally important.

The utility of the Choquet integral will be illustrated through a concise comparison in this subsection between the aggregation of the Choquet integral (CI), geometric mean (GM), and arithmetic mean (AM) methods. The outcomes of aggregating three dimensions (Economic Development (ECOD), Social Development (SOCD), and Environmental Development (ENVD)) to derive the Sustainable Development (SD) score for twelve countries are presented in Table 6.4 (for country composite SDGAI scores, refer to Table 6.5 in Appendix 6.2).[1] In order to illustrate how the composite

[1] The composite SD scores for each Choquet integral aggregation case are obtained as follows: $\mu(\{ECOD\}) * ECOD + \mu(\{SOCD\}) * SOCD + \mu(\{ENVD\}) * ENVD + [\mu(\{ECOD, SOCD\})$

SD scores, reflect the unbalanced performance of countries across the three dimensions, we computed the total absolute gaps between the three SD dimensions for every country. Specifically, we calculated the sum of the absolute gaps between the indices of ECOD and SOCD, ECOD and ENVD, and SOCD and ENVD. It is evident that as the total absolute disparities between the three dimensions increase, so does the degree of imbalance in the accomplishments across all three dimensions. The geometric mean and arithmetic mean produce nearly identical Sustainable Development (SD) scores for the following countries: Morocco and Rwanda, Colombia and Samoa, Fiji and Brazil, and Malaysia and Bulgaria, despite the variations in accomplishments across dimensions.

Utilizing the geometric mean or arithmetic mean method would not align with the objective of policymakers to prevent total substitution between dimensions and to promote comprehensive and equal results across dimensions. Both the geometric

Table 6.4 SD scores with CI, GM, and AM

Country	DECO	DSOC	DENV	SD			Total absolute gaps	Diff CI-GM
				CI	GM	AM		
Colombia	0.837	0.785	0.532	0.734	0.705	0.718	0.611	0.030
Brazil	0.561	0.827	0.793	0.746	0.716	0.727	0.464	0.029
Malaysia	0.778	0.850	0.600	0.760	0.735	0.743	0.355	0.025
Central African	0.594	0.438	0.768	0.609	0.585	0.600	0.349	0.024
Bulgaria	0.786	0.827	0.615	0.757	0.737	0.743	0.342	0.020
Togo	0.579	0.460	0.741	0.601	0.582	0.594	0.323	0.019
Morocco	0.723	0.763	0.599	0.706	0691	0.695	0.248	0.015
Rwanda	0.699	0.573	0.805	0.698	0.686	0.692	0.212	0.013
Albania	0.694	0.798	0.775	0.763	0.754	0.755	0.163	0.009
Afghanistan	0.606	0.430	0.549	0.532	0.523	0.528	0.115	0.008
Fiji	0.725	0.767	0.672	0.728	0.720	0.721	0.107	0.008
Samoa	0.677	0.757	0.691	0.715	0.707	0.708	0.029	0.008

– μ({ECOD}) – μ({SOCD})] * min(ECOD, SOCD) + [μ({ECOD, ENVD}) – μ({ECOD}) – μ({ENVD})] * min(ECOD, ENVD) + [μ({SOCD, ENVD}) – μ({SOCD}) – μ({ENVD})] * min(SOCD, ENVD) + [μ({ECOD, SOCD, ENVD}) – μ({ECOD, SOCD}) – μ({ECOD, ENVD}) – μ({SOCD, ENVD}) + μ({ECOD}) + μ({SOCD}) + μ({ENVD})] * min(ECOD, SOCD, ENVD). For example, the SD score for Colombia is obtained as follows: (0.345 * 0.837) + (0.426 * 0.785) + (0.364 * 0.532) + ((0728 – 0.426 – 0.345) * min (0.837, 0.785)) + ((0.688 – 0.345 – 0.364) * min (0.837, 0.532) + ((0.733 – 0.426 – 0.364) * min (0.785 – 0.532)) + ((1 – 0.728 – 0.688 – 0.733 + 0.364 + 0.426 + 0.345) * min (0.837 – 0.785 – 0.532)) = 0.734.

On the other hand, GM and AM obtains the aggregate SD score by:

$$\sqrt[3]{0.837 * 0.785 * 0.532} = 0.705$$

$$\frac{1}{3}(0.837 + 0.785 + 0.532) = 0.718.$$

mean and the arithmetic mean fail to consider all the potential interactions among the three dimensions of SD. However, the integral Choquet approach makes this feasible. It is evident that the absolute difference between the three dimensions is greatest for Colombia and lowest for Samoa (0.610 versus 0.028), yet their composite scores are approximately similar to those of GM and AM. However, this is evident in the aggregation of CI. The composite score derived from the CI is 0.007 greater than the composite score derived from the GM for Samoa, while this discrepancy amounts to 0.029 for Colombia. Basically, adding up the scores with the CI correctly shows the uneven results in the three areas of the overall score by considering the trade-offs between the three areas of sustainable development. In addition, the CI considers balanced achievements in two dimensions, unlike the GM. Using the standard SDG indicators and the objective weights given to all possible combinations of the indicators, it is possible to group the SDG indicators into a complete index by using the Choquet integral as an operator for combining them.

In fact, Appendix 6.2 presents the overall result of the SDG achievement index as well as its dimensions, such as the sustainable development index and the means index. In addition to the overall scores, the table gives us information on the ranking of the countries (from the most efficient to the least efficient) as well as an indication of the colours from blue to red, indicating the objectives achieved (blue) and those not yet achieved (red). While the white colour indicates that the country is halfway to achieving the target. We note that Bulgaria, followed by Romania and Georgia, are the countries closest to achieving the SDGs, while Afghanistan, Syria, and Yemen are the furthest from achieving them. This is probably due to war and political instability in these countries. We note, moreover, that most countries in sub-Saharan Africa are recording weak results, and there is still a lot of work to be done, particularly in terms of means for achieving the SDGs.

As noted, the dimension of means and strategies for achieving the Sustainable Development Goals has a considerable influence on the overall score. We observe that no country has a satisfactory level of governance. Consequently, it can be concluded that governance is a priority and an essential strategy for achieving the SDGs, both in the most disadvantaged countries and in countries with a satisfactory level of sustainable development.

Putting countries in order is not the right way to approach the 2030 Agenda because the main objective is not to get the best results. In some countries, the process of putting the SDG framework into action might be more important than the performance of the end result. Then, we use the SDG performance analysis to figure out how much work needs to be done at the national level to reach the goals. So, it is up to each state to set its own goals and choose the best indicators to measure them. To do this, we start by looking at the relationships between the variables and the dimensions of the SDG achievement index. This helps us figure out which dimensions and variables will help us reach the SDGs faster.

Figure 6.1 shows that the SDG achievement index is more positively correlated with means than with sustainable development. This means that the achievement of the SDGs essentially depends on the provision of means and strategies for sustainable development. This assertion is confirmed by the strong and positive correlation of the

POLI, GOVI, and FHCI variables with the SDG achievement index (SDGAI). Thus, there is a positive relationship between social development and the SDG achievement index. In general, the majority of the variables have a strong and positive correlation with the SDGAI, except for the environmental variables. development dimension, which has a weak and even negative correlation.

In addition to the correlation table, the scatter plot in Fig. 6.2 shows the relationship between the sustainability index and the means of implementation index. It is clear that the relationship between these two variables is linear. Following a better level of the sustainability index is a higher level of the means of implementation index. Thus, we observe that there is a high concentration of our observations below the threshold of a satisfactory level of sustainable development (0.75) and means of implementation (0.6). On the other hand, Bulgaria, Malaysia, and Romania are the countries that exceeded the thresholds of 0.75 and 0.6, respectively, of the sustainable development index and the means of implementation index, respectively.

In the same sense, it seems that the relationship between the sustainable development index and sustainable development policies is linear. Figure 6.3 (Appendix 6.2) shows that countries that have adopted favourable policies for sustainable development are the closest to achieving the goal of sustainable development. Moreover, the linear relationship is confirmed between the sustainable development index and governance, on the one hand, and between the sustainable development index and

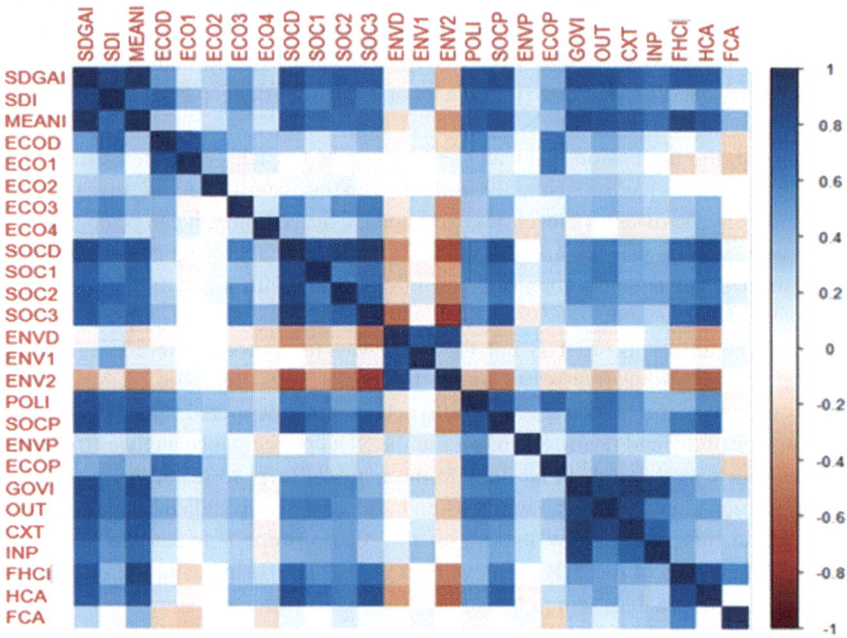

Fig. 6.1 Matrices of correlations between the SDG achievement index and their dimensions and sub-dimensions

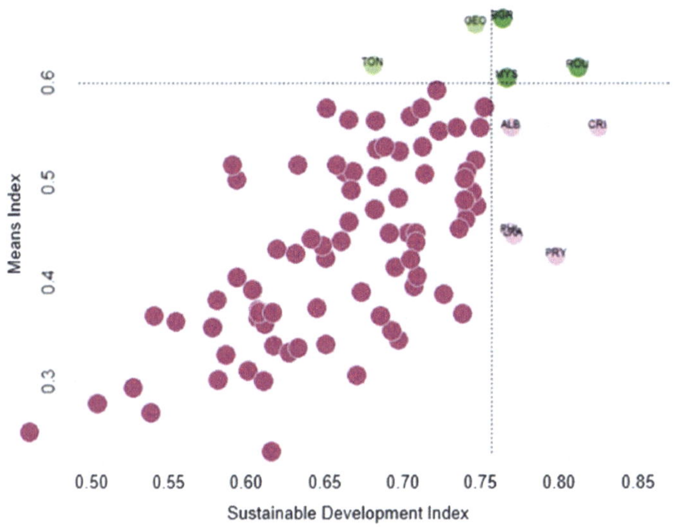

Fig. 6.2 The relationship between sustainable development and means

the capacity-building index, on the other hand. Figure 6.4 shows that governance has a positive relationship with the end goals index.

In the same way, to clearly determine the challenges and priorities of each country, Fig. 6.5 groups together, on the right, the score of each indicator of sustainable development and the dimension of the means of implementation on the left and of each country.

Although defining the priorities of each country is important, the trend towards achieving the SDGs would be useful. Therefore, we recalculated the Sustainable Development Achievement Index for five-year periods (1999–2004). In addition, Figs. 6.6, 6.7, and 6.8 illustrate the different changes in the SDGAI, SDI, and MEANI indices, respectively, between the periods 1999–2004 and 2014–2018.

Figures 6.6, 6.7, and 6.8 show the evolution of the SDGAI, SDI, and MEANI index between 1999 (blue dot) and 2018 (red dot). Regarding the SDGAI index, we notice that Georgia, Fiji, Indonesia, and Ethiopia record the most acceleration towards the achievement of the SDGs. In addition, Syria, Yemen, and Egypt have experienced a decline, which is perhaps explained by the political instability. While Ethiopia, Cambodia, and the Philippines are the countries that have made the best progress towards achieving the end goals of sustainable development, On the other hand, Ukraine, Syria, and Yemen are on the wrong track to achieve the end goals of sustainable development. Regarding the means index, Georgia has made a huge move towards achieving the mean implementation target, while Yemen, Syria, and Angola are going in the opposite direction of achieving the means objectives.

6.5 Conclusion

In this analysis, we implement our SDG assessment model. The outcome of the assessment process clearly emphasizes the integrated, indivisible, and complex nature of the SDGs. The overall SDG achievement score expressed in terms of SDGAI is a rational, transparent, and affirmative quantitative measure. In addition, our model shows a certain potential to define the strong points, the weak points, and the priorities of each country in order to follow reliable and effective policies that make it possible to achieve the SDGs by 2030. The main objective of this work is to measure progress towards the achievement of the SDGs at the level of 93 middle- and low-income countries. Indeed, our contribution is twofold. Using Keeney's (1992) value theory, we presented a new structuring of the SDGs that distinguishes between end goals and means goals. In addition, we used the Choquet integral to look for ways that the SDGs might work well together. We also used the entropy information theory to figure out how much each criterion was worth.

According to the results of the methods of construction of the index of achievement of the SDGs (SDGAI), we observed that the countries of Central Europe and Latin America are the closest to achieving the objectives of sustainable development. On the other hand, the countries of sub-Saharan Africa are the furthest away. In addition, the weak results in Yemen, Syria, and Afghanistan show that it is clear that political instability, conflicts, and wars are overriding factors that prevent the achievement of the Sustainable Development Goals. In addition, the SDGAI is composed of the Sustainable Development Index (goals of the ends of sustainable development) and the Means Index (goals of the means of the sustainable development goals); the latter was found to be the most important factor in the construction of the SDGAI. Indeed, an increase in the index of means of sustainable development leads to a considerable improvement in the SDGAI. Knowing that the sustainable development index (SDI) is composed of three dimensions: economic development (ECOD), social development (SOCD), and environmental development (ENVD), our results showed that the social pillar is the most important in the aggregation of the sustainable development index. So, we can conclude that social development is a priority to improve the results of the achievement of sustainable development goals. Moreover, the results showed us that most countries in sub-Saharan Africa have a low level of three social development indices (human capability (SOC1), foundations of well-being (SOC2), and basic human needs (SOC3)). This may be the factor that explains their lag in the Sustainable Development Achievement Index. As regards the means index, it is composed of three indices, such as policies for sustainable development (POLI), governance (GOVI), and state capacity building (FHCI). Our study has shown that strengthening financial and human capacity is the most important factor that allows an improvement in the index of means and, consequently, the index of achievement of sustainable development. In addition, we observe that there is a lot of work to be done at the level of governance in all the countries in our sample, and in particular, in the countries of sub-Saharan Africa. In order to accelerate the index of achievement of the necessary objectives, it is important to achieve considerable advancement at the level of social development, policies, and especially governance. In addition, our study not only presents the score or ranking of countries, but also presents an analytical framework to determine and analyze the gaps and priorities of each country in order to accelerate the achievement of the SDGs, which are unlikely to be achievable in Sub-Saharan African countries in 2030.

The SDG progress indices have a significant impact by providing governments with information on how they might enhance the well-being of their population while also protecting the environment. The primary objective of the United Nations was to foster equitable progress in every aspect of sustainable development. This document presents a novel analytical framework that enables the measurement of SDG progress, considering potential trade-offs between different dimensions. It also facilitates the analysis of cause-and-effect relationships and interactions among various SDGs. Additionally, it identifies strengths and weaknesses to inform the development of action plans aimed at expediting the achievement of the SDGs.

Appendix 6.1

6.6 Choquet Integral as an Aggregation Operator

After presenting the benefits of using the Choquet integral, we will now explore the essential indices that can be used to understand its behaviour in the context of an aggregate.

The Choquet integral is characterized as a measure reliant on various factors termed monotonic measures. This aggregation operator is fully employed by each of the aggregate nodes.

Let $\{a_1, a_2 \ldots a_d\}$ be the normalised criteria values of the sustainable development goals described by a set $D = \{1, 2, \ldots, d)$ of dimensions.

The Choquet Capacity, also referred to as fuzzy measurement, serves as a modelling tool for understanding the interactions among attributes. The set encompassing all these powers of M is represented as $P(M) = A[1, \ldots, A_m]$. Within the distinct functions on M lies the discrete Choquet Capacity. For a set $P(M)$ in the range [0, 1] meeting the stipulated requirements (Sugeno, 1985):

(i) $\mu(\emptyset) = 0$; $\mu(M) = 1$
(ii) For everything $S, T \subseteq M, S \subseteq T \Rightarrow \mu(S) \leq \mu(T)$

We also say that the Capacity is

- additive whenever $\mu(S \cup T) = \mu(S) + \mu(T)$ for all $S, T \subseteq M$ such that $S \cap t = \emptyset$.
- Sub additive whenever $\mu(S \cup T) \leq \mu(S) + \mu(T)$ for all $S, T \subseteq N$ such that $S \cap T = \emptyset$.
- Sub-modular whenever $\mu(S \cup T) + \mu(S \cap T) \leq \mu(S) + \mu(T)$ for all $S, T \subseteq N$.

Where S and T represent two subsets. Since the measure must be defined for each subset, if n is the number of criteria, it will have 2^n fuzzy measures.

When all criteria in the set are fully satisfied (i.e., when their normalized values are 1), and all requirements not belonging to the subset are unsatisfied, the value $\mu(S)$ can be interpreted as the weight or significance of S in the aggregation framework. This interpretation holds for each subset of characteristics $S \subseteq M$. In this context, the value $\mu(S)$ signifies the weight or importance of S when their normalized values are 0. These initial two limitations represent two boundary conditions, with the second constraint reflecting the monotonicity condition, a common requirement for any aggregation operator.

The discrete Choquet integral serves as a suitable aggregation operator that generalizes the weighted arithmetic mean. It is particularly useful in situations where a Capacity is employed to denote the relevance of subsets of attributes. As per Grabisch (1996) and Grabisch et al. (2008), the Choquet integral of a function a, denoted by MR in relation to a Capacity on M, can be described using the following formula:

$$C_\mu(a) := \sum_{i=1}^{m} a_{(i)}\left[\mu\left(B_{(i)}\right) - \mu(B_{(i+1)})\right]$$

where the notation (.) indicates a permutation such that $a_1 \leq \cdots \leq a_m$, $B_{(i)} := \left\{A_{(i)}, \ldots, A_{(m)}\right\}$, for everything $I \in \{1, \ldots, m\}$, and $B_{(m+1)} := \emptyset$.

Viewed as an aggregation operator, the Choquet integral with respect to μ takes into consideration the interaction phenomena among attributes, encompassing the effects of complementarity or redundancy between elements of M, which are modelled by μ. Specifically, complementarity (or redundancy) between two disjoint subsets of attributes A and B is naturally captured by the inequality $\mu(A \cup B) \geq \mu(A) + \mu(B)$ (or $\leq$). It's noteworthy that if μ is a submodular Capacity, it can only model redundant interactions between two disjoint subsets of attributes.

To gauge the overall importance of an attribute A_i, its Shapley value (Shapley, 1953) can be utilized, as expressed by

$$\Phi(\mu, A_i) := \sum_{T \subseteq M \setminus \{A_i\}} \frac{(m - |T| - 1)!|T|!}{m!}[\mu(T \cup \{A_i\}) - \mu(T)]$$

The Shapley value of an attribute A_i can be considered as an average value of the marginal contribution $\mu(T \cup \{A_i\}) - \mu(T)$ of the attribute A_i to a subset T which does not contain it.

The average interaction (Grabisch, 1997) between two attributes A_i and A_j can be measured using their Shapley interaction index, which can be calculated using:

$$I\left(\mu, \left\{A_i, A_j\right\}\right) := \sum_{T \subseteq M \setminus \left\{A_i, A_j\right\}} \frac{(m - |T| - 2)!|T|}{(m - 1)!}$$
$$* \left[\mu\left(T \cup \left\{A_i, A_j\right\}\right) - \mu(T \cup \{A_i\}) - \mu(T \cup \left\{A_j\right\} + \mu(T)\right]$$

The primary data available in the context of aggregation using the Choquet integral, particularly when no baseline preferences are known, is the set of profiles. This set becomes instrumental in determining the weights of attribute subsets (i.e., the underlying Capacity). Recognizing this, Grabisch et al. (2008) introduced a measure-induced entropy as an unsupervised approach to Capacity identification. This entropy serves to estimate the interaction between characteristics. The proposed method primarily involves replacing the notion of the subjective relevance of a subset of qualities with the probabilistic concept of the informative content of a subset of attributes, a computable quantity based on the set profiles.

The concept of entropy as a measure of information contained in a random sample was initially introduced by Shannon (1948). It gauges uncertainty, information, or the structure of probability distribution. The Shannon entropy of a probability distribution p specified on a finite and non-empty set N can be defined as follows:

$$H_S(p) := \sum_{i \in N} h\big[p(i)\big]$$

When: $h(x) = \begin{cases} -xLnx, & x > 0 \\ 0, & si\ x = 0 \end{cases}.$

A stochastic system's level of uncertainty is measured by $H_S(p)$. It depicts the unpredictable nature of that system in a probabilistic setting (Cover, 1987). If and only if p is uniform, it achieves its maximum value (ln N) (property of maximality).

The weight of each subset $S \subseteq M$ of attributes can then be defined by following formula. We assume that the random variables $X_1, \ldots., X_m$ are discrete and take their values from the finite sets $\aleph_1, \ldots, \aleph_m$. To ensure that the resulting set function is monotonic, we must first assume that they are discrete.

$$\mu(S) := \begin{cases} 0, & s = 0 \\ \frac{h\{x1.....xs\}}{h(n)}, & si\ S = \{A1 \ldots\ldots.Ais\} \end{cases}$$

where h is the generic set function defined by the previous equation. It should be noted that is well-defined if the underlying entropy measure satisfies the decision property. Indeed, h($\aleph$) $\neq$ 0 in this case because it was assumed that $p\ (X_1, \ldots, X_m)$ is not a Dirac mass.

An optimization model is developed to obtain the weight of each criterion and the interaction between the criteria. The weighting of the criteria will then be used to calculate the overall satisfaction of the alternative solutions by aggregation with the evaluation scores. With the contribution obtained objectively from the criteria, an optimization model has been developed to maximize Marichal's entropy. The model is given as:

$$max_{\mu} H_M(\mu) = -\sum_{i=1}^{m} \sum_{T \subseteq M \setminus \{A_I\}} \frac{(m - |T| - 1)!|T|!}{m!}$$

$$* \, [\mu(T \cup \{A_i\}) - \mu(T)] ln[\mu(T \cup \{A_i\}) - \mu(T)]$$

$$\text{s.c} \begin{cases} \Phi(\mu, A_i) := \sum_{T \subseteq M \setminus \{A_I\}} \frac{(m-|T|-1)!|T|!}{m!} [\mu(T \cup \{A_i\}) - \mu(T)] \ \ i = 1, \ldots, m \\ \mu(A) = 1 \\ \mu(A_i) \in [0, 1], \forall A_i \in P(A) \end{cases} .$$

Appendix 6.2

6.7 SDG Achievement Index: An analytical framework

Table 6.5 The score of the SDG achievement index and its components for the period 2014–2018

Country	SDGAI			Sustainable development			Means		
	SDGAI	SD	MEAN	DECO	DSOC	DENV	POL	GOV	FHC
Bulgaria	0.7049	0.7571	0.6658	0.7863	0.8269	0.6154	0.8144	0.4892	0.6856
Romania	0.6976	0.8057	0.6164	0.8503	0.7979	0.7602	0.7472	0.4900	0.6120
Georgia	0.6940	0.7400	0.6594	0.7496	0.7654	0.6895	0.6134	0.5298	0.7756
Malaysia	0.6723	0.7601	0.6064	0.7775	0.8502	0.5999	0.7367	0.5259	0.5680
Costa Rica	0.6686	0.8185	0.5560	0.7372	0.8846	0.7975	0.5494	0.5835	0.5326
Brazil	0.6486	0.7455	0.5758	0.5610	0.8267	0.7929	0.5840	0.4149	0.6808
Samoa	0.6454	0.7151	0.5930	0.6766	0.7567	0.6909	0.5420	0.5799	0.6298
Albania	0.6450	0.7630	0.5563	0.6935	0.7979	0.7748	0.6384	0.4504	0.5736
Tonga	0.6426	0.6746	0.6187	0.6503	0.7291	0.6161	0.4862	0.4735	0.7975
Montenegro	0.6358	0.7426	0.5556	0.7022	0.7810	0.7243	0.6101	0.4623	0.5844
Serbia	0.6308	0.7049	0.5751	0.6248	0.8344	0.5914	0.5639	0.4486	0.6683
Fiji	0.6297	0.7281	0.5557	0.7248	0.7667	0.6715	0.5425	0.4447	0.6395
Belarus	0.6245	0.6990	0.5686	0.5504	0.8385	0.6348	0.6718	0.2921	0.6953
Bosnia and Herzegovina	0.6234	0.6981	0.5673	0.6865	0.8014	0.5493	0.6326	0.3722	0.6614
Armenia	0.6228	0.7168	0.5521	0.6997	0.7931	0.6166	0.5495	0.3754	0.6745
Thailand	0.6161	0.7399	0.5231	0.7889	0.7955	0.5987	0.7531	0.3735	0.4611
Azerbaijan	0.6111	0.6762	0.5622	0.6310	0.7276	0.6441	0.5210	0.2771	0.7817
Morocco	0.6091	0.7060	0.5363	0.7230	0.7625	0.5989	0.7027	0.3649	0.5387
Turkey	0.6072	0.7345	0.5116	0.7278	0.8185	0.6105	0.6363	0.3646	0.5273
Namibia	0.6051	0.6445	0.5755	0.5765	0.6112	0.7282	0.5955	0.5114	0.6073
Jordan	0.6048	0.6590	0.5641	0.5788	0.8163	0.5018	0.5343	0.4129	0.6854
Colombia	0.6047	0.7342	0.5074	0.8374	0.7850	0.5320	0.7356	0.4018	0.4139
Viet Nam	0.6022	0.7325	0.5043	0.8246	0.7408	0.6043	0.6430	0.3420	0.5210
Jamaica	0.6001	0.6912	0.5317	0.5814	0.8118	0.6195	0.4694	0.4811	0.6029
Vanuatu	0.5988	0.6814	0.5368	0.6341	0.6912	0.7061	0.4868	0.4684	0.6137
Indonesia	0.5969	0.7381	0.4909	0.7486	0.7344	0.7300	0.5909	0.3921	0.4892
Russia	0.5953	0.6763	0.5345	0.5844	0.8015	0.5824	0.6690	0.2795	0.6239
Suriname	0.5942	0.7070	0.5094	0.5799	0.8031	0.6842	0.4845	0.4157	0.5885
El Salvador	0.5903	0.7410	0.4772	0.6918	0.8047	0.6954	0.5712	0.4066	0.4589
Peru	0.5898	0.7329	0.4824	0.6769	0.8053	0.6813	0.5423	0.4206	0.4834
Argentina	0.5887	0.7336	0.4799	0.5837	0.8843	0.6549	0.5354	0.4106	0.4898
Philippines	0.5843	0.7622	0.4507	0.8278	0.7572	0.6861	0.5819	0.3796	0.4048
Sri Lanka	0.5834	0.7640	0.4478	0.7604	0.7738	0.7527	0.5639	0.4024	0.3930
Paraguay	0.5834	0.7911	0.4274	0.7274	0.7827	0.8463	0.4917	0.3480	0.4383
Tunisia	0.5794	0.6763	0.5066	0.6016	0.7835	0.5916	0.5906	0.3817	0.5366
Gabon	0.5792	0.7336	0.4633	0.6222	0.7070	0.8424	0.5512	0.2907	0.5260
Ghana	0.5755	0.6616	0.5109	0.5742	0.6503	0.7372	0.4950	0.4449	0.5654

(continued)

Table 6.5 (continued)

Country	SDGAI			Sustainable development			Means		
	SDGAI	SD	MEAN	DECO	DSOC	DENV	POL	GOV	FHC
Kyrgyzstan	0.5752	0.6509	0.5183	0.6142	0.7405	0.5513	0.5476	0.2794	0.6639
Belize	0.5735	0.6561	0.5115	0.5703	0.7420	0.6114	0.4682	0.3913	0.6197
Guyana	0.5734	0.6904	0.4855	0.6406	0.7370	0.6692	0.5187	0.3829	0.5356
Nicaragua	0.5726	0.7296	0.4547	0.7061	0.7019	0.7743	0.5905	0.3135	0.4578
Kazakhstan	0.5647	0.6600	0.4931	0.6403	0.7855	0.4856	0.5179	0.3464	0.5784
Mongolia	0.5645	0.6261	0.5183	0.5885	0.7543	0.4671	0.6506	0.4416	0.4757
Senegal	0.5600	0.6753	0.4734	0.7112	0.6230	0.6867	0.4947	0.4192	0.4977
Ecuador	0.5580	0.7018	0.4501	0.5861	0.8196	0.6388	0.5234	0.3206	0.4916
Bolivia	0.5557	0.6963	0.4501	0.6804	0.7440	0.6395	0.5027	0.3141	0.5113
Honduras	0.5529	0.7017	0.4411	0.6580	0.7214	0.7123	0.5389	0.3008	0.4722
Tajikistan	0.5505	0.6845	0.4500	0.7386	0.6757	0.6283	0.4817	0.1781	0.6166
Ukraine	0.5468	0.5843	0.5186	0.4046	0.7695	0.4837	0.4533	0.2701	0.7277
Lebanon	0.5458	0.6590	0.4608	0.5768	0.7690	0.5779	0.4049	0.2660	0.6275
Rwanda	0.5411	0.6982	0.4232	0.6990	0.5725	0.8052	0.5004	0.4260	0.3564
South Africa	0.5393	0.5868	0.5036	0.4997	0.7696	0.3975	0.5072	0.4838	0.5148
Laos	0.5332	0.7022	0.4064	0.7840	0.5683	0.7416	0.4757	0.2805	0.4482
Comoros	0.5324	0.6877	0.4158	0.6771	0.6575	0.7231	0.5268	0.2784	0.4347
Guatemala	0.5303	0.7194	0.3882	0.6772	0.6954	0.7747	0.5016	0.3130	0.3588
Cambodia	0.5263	0.7004	0.3956	0.8234	0.4532	0.8016	0.5923	0.2841	0.3304
Uzbekistan	0.5255	0.6345	0.4436	0.7268	0.7118	0.4087	0.5781	0.1927	0.5298
Nepal	0.5250	0.6415	0.4375	0.5815	0.6501	0.6769	0.4554	0.2778	0.5357
Myanmar	0.5246	0.7317	0.3692	0.8567	0.5331	0.7867	0.5221	0.2021	0.3787
China	0.5218	0.6290	0.4413	0.6195	0.5752	0.6837	0.4959	0.3436	0.4731
India	0.5185	0.6438	0.4244	0.8517	0.6155	0.4204	0.5732	0.3952	0.3276
Zambia	0.5130	0.6250	0.4289	0.4492	0.4839	0.8974	0.4971	0.3721	0.4200
Mozambique	0.5101	0.6124	0.4332	0.4747	0.4872	0.8407	0.5748	0.2829	0.4387
Venezuela	0.5092	0.6668	0.3909	0.4783	0.8119	0.6294	0.4840	0.1278	0.5145
Kenya	0.5002	0.6791	0.3659	0.6096	0.5980	0.8107	0.4204	0.3163	0.3618
Cameroon	0.4951	0.6858	0.3519	0.8074	0.5994	0.6424	0.4974	0.2160	0.3437
Bangladesh	0.4913	0.6899	0.3421	0.8487	0.6343	0.5671	0.5378	0.2484	0.2643
cote d'ivoire	0.4878	0.6385	0.3746	0.5729	0.3844	0.9152	0.4539	0.2951	0.3741
Burkina Faso	0.4835	0.5871	0.4058	0.6137	0.3327	0.7816	0.4918	0.3371	0.3921
Egypt	0.4803	0.5971	0.3926	0.5802	0.7570	0.3657	0.4524	0.2217	0.4734
Zimbabwe	0.4731	0.6101	0.3703	0.5349	0.5671	0.7089	0.4683	0.1593	0.4542
Liberia	0.4704	0.6006	0.3726	0.4247	0.4801	0.8520	0.2929	0.2714	0.4897
Mali	0.4694	0.6439	0.3383	0.7484	0.4043	0.7563	0.4407	0.2513	0.3261
Togo	0.4688	0.6010	0.3695	0.5794	0.4602	0.7411	0.4397	0.2706	0.3904
Gambia	0.4652	0.6005	0.3637	0.5140	0.5048	0.7597	0.4012	0.2814	0.3970
Benin	0.4644	0.5740	0.3821	0.5955	0.4429	0.6672	0.4715	0.3740	0.3154
Nigeria	0.4637	0.6046	0.3579	0.5801	0.5059	0.7111	0.4304	0.1978	0.4230
Ethiopia	0.4598	0.6639	0.3066	0.6808	0.4200	0.8580	0.4321	0.2220	0.2750

(continued)

Table 6.5 (continued)

Country	SDGAI			Sustainable development			Means		
	SDGAI	SD	MEAN	DECO	DSOC	DENV	POL	GOV	FHC
Malawi	0,4598	0,6263	0,3347	0,4601	0,5605	0,8151	0,3955	0,3455	0,2742
Guinea	0,4547	0,6207	0,3299	0,6249	0,3781	0,8253	0,4355	0,2278	0,3266
Angola	0,4540	0,6107	0,3362	0,5015	0,5232	0,7798	0,3395	0,2149	0,4171
Uganda	0,4474	0,5711	0,3546	0,5274	0,4583	0,7067	0,4722	0,3079	0,2999
Sierra Leone	0,4410	0,5481	0,3606	0,3656	0,3772	0,8558	0,3895	0,2934	0,3890
Niger	0,4380	0,5334	0,3663	0,5068	0,3622	0,7041	0,4175	0,2779	0,3938
Madagascar	0,4362	0,5801	0,3281	0,5503	0,3746	0,7831	0,4059	0,2770	0,3077
Haiti	0,4332	0,5944	0,3121	0,5456	0,3900	0,8133	0,2947	0,1963	0,4017
Pakistan	0,4310	0,6043	0,3009	0,6757	0,5672	0,5661	0,4493	0,2081	0,2581
Burundi	0,4196	0,5751	0,3028	0,4129	0,4394	0,8321	0,4390	0,1460	0,3172
Central African	0,3925	0,6087	0,2302	0,5941	0,4377	0,7684	0,4240	0,0865	0,1913
Chad	0,3912	0,5197	0,2947	0,5288	0,2662	0,7290	0,5206	0,1320	0,2457
Afghanistan	0,3819	0,5315	0,2695	0,6062	0,4303	0,5486	0,2354	0,0906	0,4121
Syria	0,3725	0,4972	0,2789	0,4158	0,6627	0,3286	0,3315	0,0125	0,4293
Yemen	0,3373	0,4535	0,2501	0,4755	0,4218	0,4601	0,2873	0,0513	0,3635

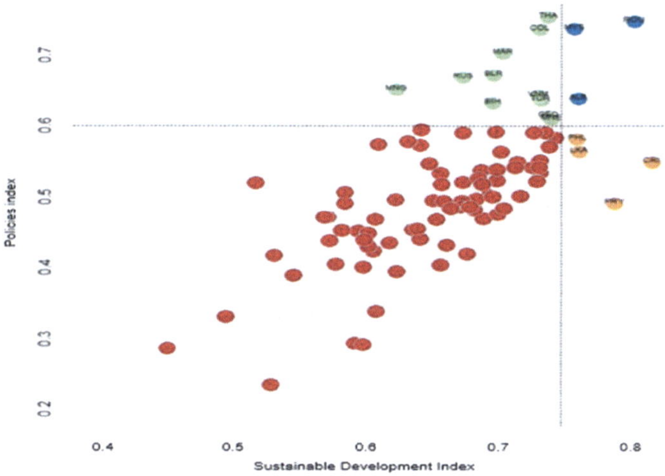

Fig. 6.3 The relationship between SD and policies

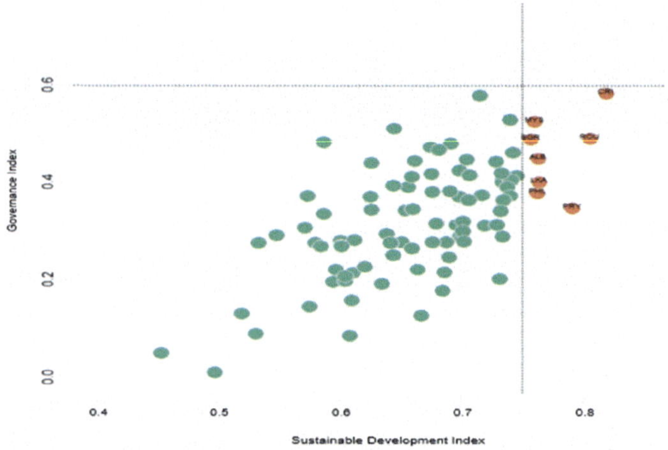

Fig. 6.4 The relationship between SD and governance

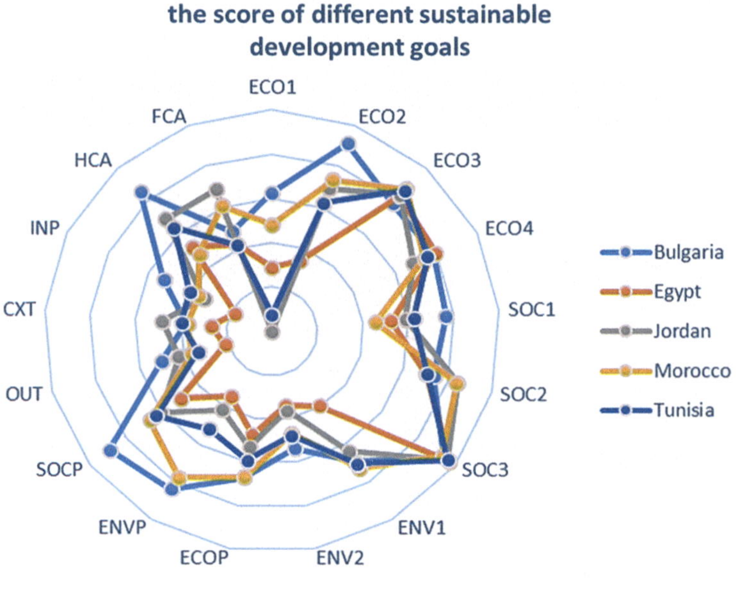

Fig. 6.5 The score of different sustainable development goals

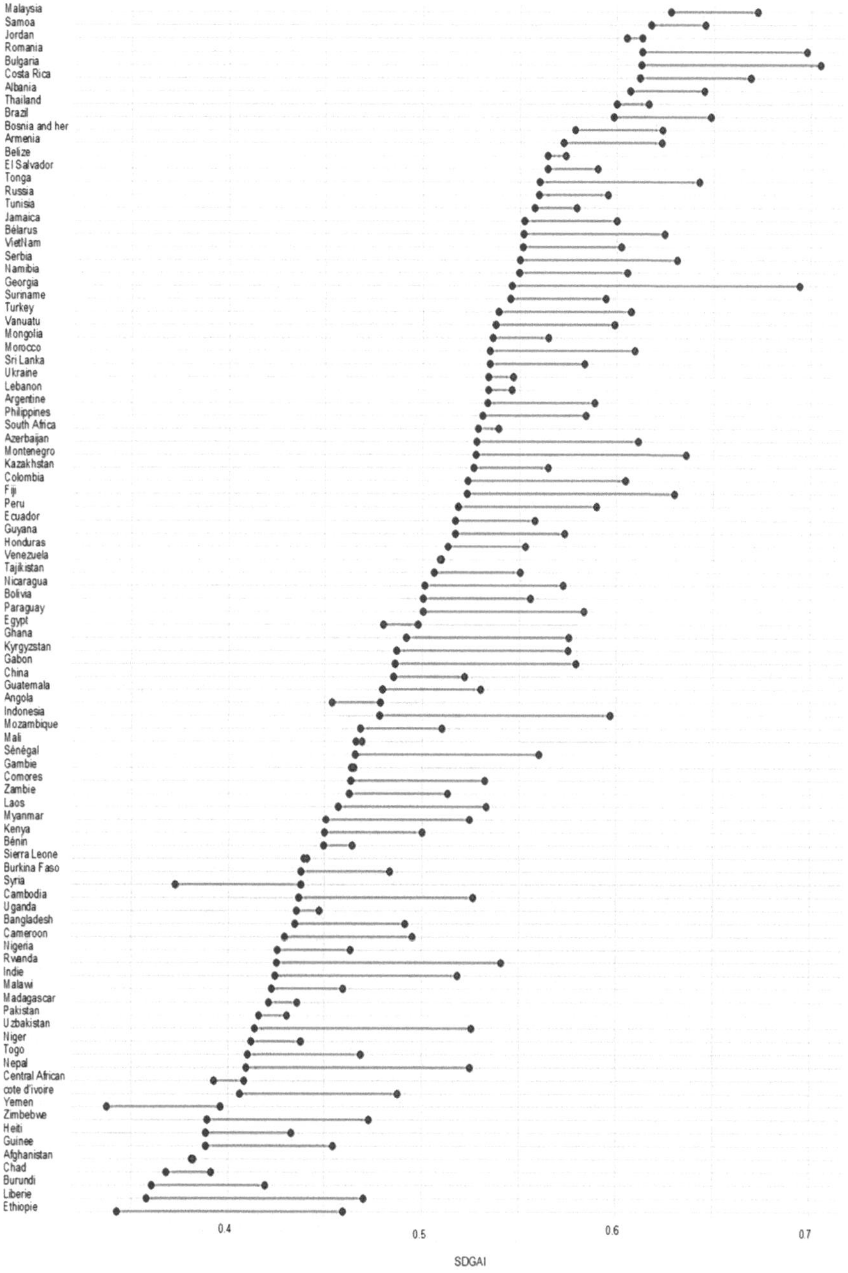

Fig. 6.6 The change in the SDGAI

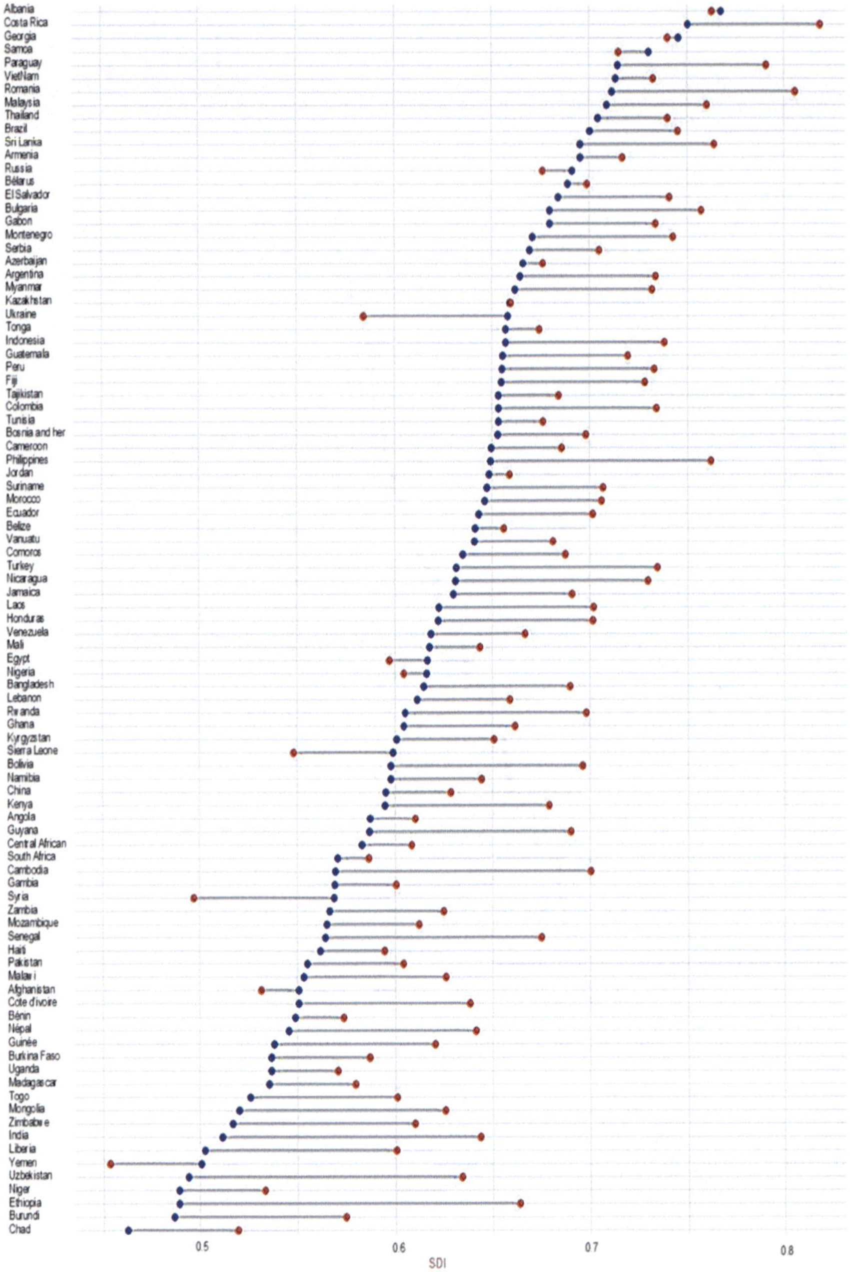

Fig. 6.7 The change in the SD index

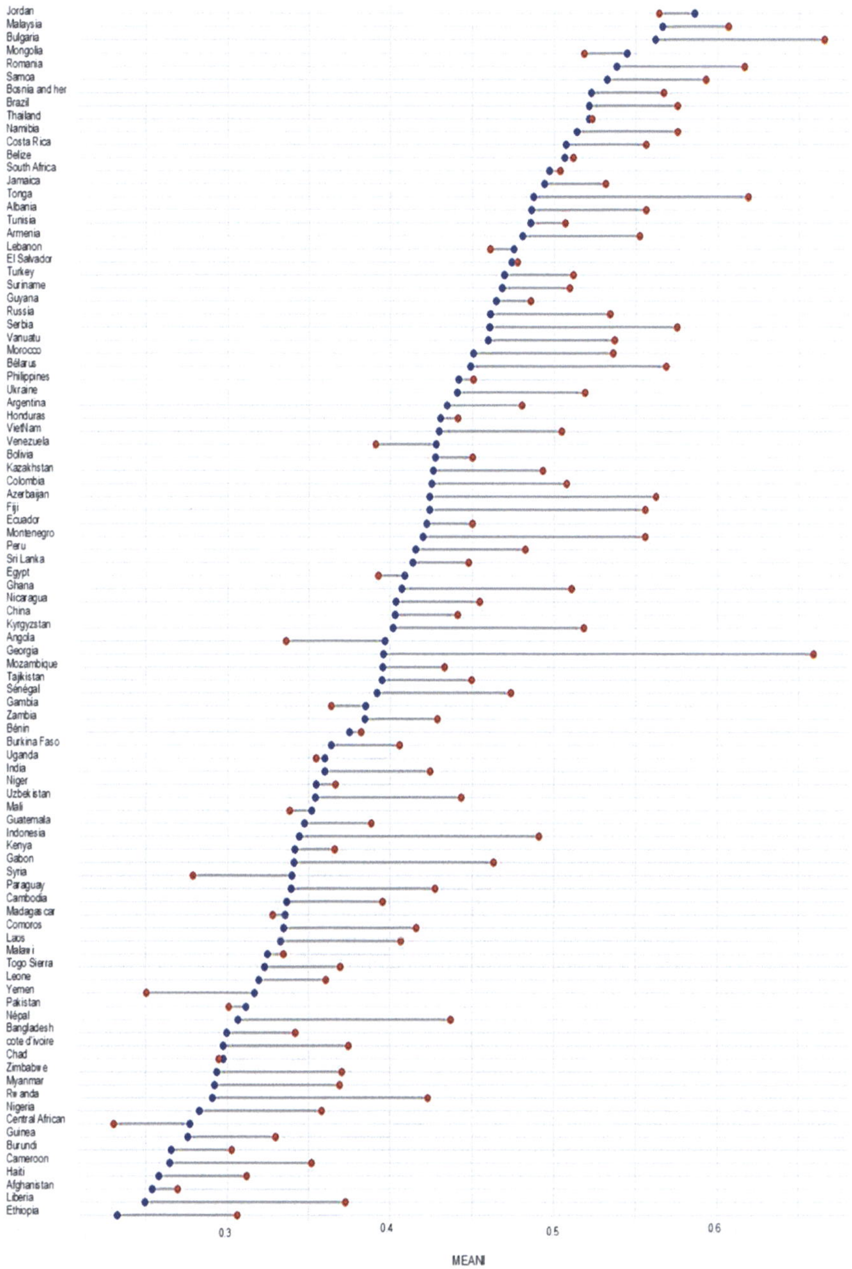

Fig. 6.8 The change in the MEANS index

References

Allen, C., Metternicht, G., & Wiedmann, T. (2021). Priorities for science to support national implementation of the sustainable development goals: A review of progress and gaps. *Sustainable Development, 29*(4), 635–652. https://doi.org/10.1002/sd.2164

Asadikia, A., Rajabifard, A., & Kalantari, M. (2021). Systematic prioritisation of SDGs: Machine learning approach. *World Development, 140*, 105269. https://doi.org/10.1016/j.worlddev.2020.105269

Bali Swain, R., & Ranganathan, S. (2021). Modeling interlinkages between sustainable development goals using network analysis. *World Development, 138*, 105136. https://doi.org/10.1016/j.worlddev.2020.105136

Barbier, E. B., & Burgess, J. C. (2019). Sustainable development goal indicators: Analyzing tradeoffs and complementarities. *World Development, 122*, 295–305. https://doi.org/10.1016/j.worlddev.2019.05.026

Bartniczak, B., & Raszkowski, A. (2018). Sustainable development in African countries: An indicator-based approach and recommendations for the future. *Sustainability, 11*(1), 22. https://doi.org/10.3390/su11010022

Bennich, T., Persson, S., Beaussart, R., Allen, C., & Malekpour, S. (2023). Recurring patterns of SDG interlinkages and how they can advance the 2030 agenda. *One Earth, 6*(11), 1465–1476. https://doi.org/10.1016/j.oneear.2023.10.008

Biggeri, M., & Mauro, V. (2018). Towards a more 'Sustainable' human development index: Integrating the environment and freedom. *Ecological Indicators, 91*, 220–231. https://doi.org/10.1016/j.ecolind.2018.03.045

Biggeri, M., Clark, D. A., Ferrannini, A., & Mauro, V. (2019). Tracking the SDGs in an 'integrated' manner: A proposal for a new index to capture synergies and trade-offs between and within goals. *World Development, 122*, 628–647. https://doi.org/10.1016/j.worlddev.2019.05.022

Brelsford, C., Lobo, J., Hand, J., & Bettencourt, L. M. A. (2017). Heterogeneity and scale of sustainable development in cities. *Proceedings of the National Academy of Sciences, 114*(34), 8963–8968. https://doi.org/10.1073/pnas.1606033114

Breuer, A., Leininger, J., Malerba, D., & Tosun, J. (2023). Integrated policymaking: Institutional designs for implementing the sustainable development goals (SDGs). *World Development, 170*, 106317. https://doi.org/10.1016/j.worlddev.2023.106317

Chen, Y., Liu, C., Li, H., & Xue, X. (2023). How do countries along the Maritime Silk Road perform in sustainable use of natural resources? Progress of natural resources-related SDGs. *Ecological Indicators, 149*, 110194. https://doi.org/10.1016/j.ecolind.2023.110194

Choquet, G. (1954). Theory of capacity. *Annales De L'institut Fourier, 5*, 131–295. https://doi.org/10.5802/aif.53

Clark, D. A. (2005). Sen's capability approach and the many spaces of human well-being. *Journal of Development Studies, 41*(8), 1339–1368. https://doi.org/10.1080/00220380500186853

Clark, D. A. (2016). Valuing and revaluing education: What can we learn about measurement from the South African poor? *Comparative Education, 53*(1), 54–80. https://doi.org/10.1080/03050068.2017.1254954

Cling, J. P., & Delecourt, C. (2022). Interlinkages between the sustainable development goals. *World Development Perspectives, 25*, 100398. https://doi.org/10.1016/j.wdp.2022.100398

Coenen, J., Glass, L. M., & Sanderink, L. (2021). Two degrees and the SDGs: A network analysis of the interlinkages between transnational climate actions and the sustainable development goals. *Sustainability Science, 17*(4), 1489–1510. https://doi.org/10.1007/s11625-021-01007-9

Cohen, E. H. (2004). Assessing the human development index through the structure of welfare. *Bulletin of Sociological Methodology/bulletin De Méthodologie Sociologique, 84*(1), 40–48. https://doi.org/10.1177/075910630408400102

Costanza, R., Daly, L., Fioramonti, L., Giovannini, E., Kubiszewski, I., Mortensen, L. F., & Wilkinson, R. (2016). Modelling and measuring sustainable wellbeing in connection with the

UN sustainable development goals. *Ecological Economics, 130*, 350–355. https://doi.org/10.1016/j.ecolecon.2016.07.009

Cover, T. M. (1987). Information theory and the stock market. *Mathematical Social Sciences, 13*(3), 298. https://doi.org/10.1016/0165-4896(87)90039-4

Dang, H. A. H., & Serajuddin, U. (2020). Tracking the sustainable development goals: Emerging measurement challenges and further reflections. *World Development, 127*, 104570. https://doi.org/10.1016/j.worlddev.2019.05.024

Dawes, J. (2022). SDG interlinkage networks: Analysis, robustness, sensitivities, and hierarchies. *World Development, 149*, 105693. https://doi.org/10.1016/j.worlddev.2021.105693

Dawes, J. H. P., Zhou, X., & Moinuddin, M. (2022). System-level consequences of synergies and trade-offs between SDGs: Quantitative analysis of interlinkage networks at country level. *Sustainability Science, 17*(4), 1435–1457. https://doi.org/10.1007/s11625-022-01109-y

Eurostat. (2021). Sustainable development in the european union: Monitoring report on progress towards the sdgs in an eu context. Luxembourg.

Fang, K., Xu, A., Wang, S., Jia, X., Liao, Z., Tan, R. R., … Su, F. (2023). Progress towards sustainable development goals in the belt and road initiative countries. *Journal of Cleaner Production, 424*, 138808. https://doi.org/10.1016/j.jclepro.2023.138808

Françozo, R. V., & Belderrain, M. C. N. (2022). A problem structuring method framework for value-focused thinking. *EURO Journal on Decision Processes, 10*, 100014. https://doi.org/10.1016/j.ejdp.2022.100014

Gini, C. (1921). Measurement of inequality of incomes. *The Economic Journal, 31*(121), 124. https://doi.org/10.2307/2223319

Giupponi, C., & Gain, A. K. (2016). Integrated spatial assessment of the water, energy and food dimensions of the sustainable development goals. *Regional Environmental Change, 17*(7), 1881–1893. https://doi.org/10.1007/s10113-016-0998-z

Grabisch, M. (1996). The application of fuzzy integrals in multicriteria decision making. *European Journal of Operational Research, 89*(3), 445–456. https://doi.org/10.1016/0377-2217(95)00176-x

Grabisch, M. (1997). K-order additive discrete fuzzy measures and their representation. *Fuzzy Sets and Systems, 92*(2), 167–189. https://doi.org/10.1016/s0165-0114(97)00168-1

Grabisch, M., Kojadinovic, I., & Meyer, P. (2008). A review of methods for capacity identification in choquet integral based multi-attribute utility theory. *European Journal of Operational Research, 186*(2), 766–785. https://doi.org/10.1016/j.ejor.2007.02.025

Guerrero, O. A., Guariso, D., & Castañeda, G. (2023). Aid effectiveness in sustainable development: A multidimensional approach. *World Development, 168*, 106256. https://doi.org/10.1016/j.worlddev.2023.106256

Guo, H., Huang, L., Luo, L., Liu, J., & Li, X. (2023). Progress on achieving environmental SDGs assessed from big earth data in China. *Science Bulletin*. https://doi.org/10.1016/j.scib.2023.09.017

Guo, H., Liang, D., Sun, Z., Chen, F., Wang, X., Li, J., … Shirazi, Z. (2022). Measuring and evaluating SDG indicators with big earth data. *Science Bulletin, 67*(17), 1792–1801. https://doi.org/10.1016/j.scib.2022.07.015

Haq, M. (1995). *Reflections on human development*. Oxford University Press.

Heck, V., Hoff, H., Wirsenius, S., Meyer, C., & Kreft, H. (2018). Land use options for staying within the planetary boundaries—Synergies and trade-offs between global and local sustainability goals. *Global Environmental Change, 49*, 73–84. https://doi.org/10.1016/j.gloenvcha.2018.02.004

Hicks, N., & Streeten, P. (1979). Indicators of development: The search for a basic needs yardstick. *World Development, 7*(6), 567–580.

Horn, P., & Grugel, J. (2018). The SDGs in middle-income countries: Setting or serving domestic development agendas? Evidence from Ecuador. *World Development, 109*, 73–84. https://doi.org/10.1016/j.worlddev.2018.04.005

Huang, R. (2023). SDG-oriented sustainability assessment for Central and Eastern European countries. *Environmental and Sustainability Indicators, 19*, 100268. https://doi.org/10.1016/j.indic.2023.100268

Issa, L., Mezher, T., & El Fadel, M. (2024). Can network analysis ascertain SDGs interlinkages towards evidence-based policy planning? A systematic critical assessment. *Environmental Impact Assessment Review, 104*, 107295. https://doi.org/10.1016/j.eiar.2023.107295

Ivanov, A., & Peleah, M. (2018). Sustainable humand development index—A pragmatic proposal for monitoring sustainability within the affordable boundaries November 9–11. Paper presented at the Sustainability and Development Meeting. University of Michigan.

Kanbur, R., & Stiglitz, J. E. (2016). Dynastic inequality, mobility and equality of opportunity. *The Journal of Economic Inequality, 14*(4), 419–434. https://doi.org/10.1007/s10888-016-9328-4

Keeney, R. L. (1992). *Value-focused thinking: A path to creative decisionmaking*. http://ci.nii.ac.jp/ncid/BA27833135

Keeney, R. L. (1996). Value-focused thinking: Identifying decision opportunities and creating alternatives. *European Journal of Operational Research, 92*(3), 537–549. https://doi.org/10.1016/0377-2217(96)00004-5

Kılkış, I. (2016). Sustainable development of energy, water and environment systems index for Southeast European cities. *Journal of Cleaner Production, 130*, 222–234. https://doi.org/10.1016/j.jclepro.2015.07.121

Kostetckaia, M., & Hametner, M. (2022). How sustainable development goals interlinkages influence European Union countries' progress towards the 2030 agenda. *Sustainable Development*. https://doi.org/10.1002/sd.2290

Kroll, C., Warchold, A., & Pradhan, P. (2019). Sustainable development goals (SDGs): Are we successful in turning trade-offs into synergies? *Palgrave Communications, 5*(1). https://doi.org/10.1057/s41599-019-0335-5

Kuc-Czarnecka, M., Markowicz, I., & Sompolska-Rzechuła, A. (2023). SDGs implementation, their synergies, and trade-offs in EU countries—Sensitivity analysis-based approach. *Ecological Indicators, 146*, 109888. https://doi.org/10.1016/j.ecolind.2023.109888

Kunz, R. E., Siebert, J., & Mütterlein, J. (2016). Combining value-focused thinking and balanced scorecard to improve decision-making in strategic management. *Journal of Multi-criteria Decision Analysis, 23*(5–6), 225–241. https://doi.org/10.1002/mcda.1572

Kynčlová, P., Upadhyaya, S., & Nice, T. (2020). Composite index as a measure on achieving Sustainable Development Goal 9 (SDG-9) industry-related targets: The SDG-9 index. *Applied Energy, 265*, 114755. https://doi.org/10.1016/j.apenergy.2020.114755

Le Blanc, D. (2015). Towards integration at last? The sustainable development goals as a network of targets. *Sustainable Development, 23*(3), 176–187. https://doi.org/10.1002/sd.1582

Liu, S. (2020). Interlinkages between indicators of sustainable development goals: Evidence from seven low income and lower middle-income countries. *Sustainable Development Research, 2*(1), 58. https://doi.org/10.30560/sdr.v2n1p58

Liu, F., Shang, Y. F., & Pan, L. H. (2015). A modified TOPSIS method for obtaining the associated weights of the OWA-type operators. *International Journal of Intelligent Systems, 30*(10), 1101–1116. https://doi.org/10.1002/int.21737

Lu, Z. N., Chen, H., Hao, Y., Wang, J., Song, X., & Mok, T. M. (2017). The dynamic relationship between environmental pollution, economic development and public health: Evidence from China. *Journal of Cleaner Production, 166*, 134–147. https://doi.org/10.1016/j.jclepro.2017.08.010

Mononen, L., Auvinen, A. P., Ahokumpu, A. L., Rönkä, M., Aarras, N., Tolvanen, H., … Vihervaara, P. (2016). National ecosystem service indicators: Measures of social–ecological sustainability. *Ecological Indicators, 61*, 27–37. https://doi.org/10.1016/j.ecolind.2015.03.041

Moyer, J. D., & Hedden, S. (2020). Are we on the right path to achieve the sustainable development goals? *World Development, 127*, 104749. https://doi.org/10.1016/j.worlddev.2019.104749

Nerland, R., Nilsen, H. R., & Andersen, B. (2022). Biosphere-based sustainability in local governments: Sustainable development goal interactions and indicators for policymaking. *Sustainable Development*. https://doi.org/10.1002/sd.2371

Nhemachena, C., Matchaya, G., Nhemachena, C., Karuaihe, S., Muchara, B., & Nhlengethwa, S. (2018). Measuring baseline agriculture-related sustainable development goals index for Southern Africa. *Sustainability, 10*(3), 849. https://doi.org/10.3390/su10030849

Nilsson, M., Chisholm, E., Griggs, D., Howden-Chapman, P., McCollum, D., Messerli, P., …, Stafford-Smith, M. (2018). Mapping interactions between the sustainable development goals: lessons learned and ways forward. *Sustainability Science, 13*(6), 1489–1503. https://doi.org/10.1007/s11625-018-0604-z

Obersteiner, M., Walsh, B., Frank, S., Havlík, P., Cantele, M., Liu, J., et al. (2016). Assessing the land resource–food price nexus of the sustainable development goals. *Science Advances, 2*(9), e1501499.

OECD. (2017). Measuring distance to the SDG targets: An assessment of where OECD countries stand. OECD. http://www.oecd.org/std/OECD-Measuring-Distance-to[1]SDG-Targets.pdf

Osman, A., Mensah, E. A., Mensah, C. A., Asamoah, Y., Dauda, S., Adu-Boahen, K., & Adongo, C. A. (2022). Spatial analysis of synergies and trade-offs between the sustainable development goals (SDGs) in Africa. *Geography and Sustainability, 3*(3), 220–231. https://doi.org/10.1016/j.geosus.2022.07.003

Pinar, M. (2021). Choquet-integral aggregation method to aggregate social indicators to account for interactions: An application to the human development index. *Social Indicators Research, 159*(1), 1–53. https://doi.org/10.1007/s11205-021-02726-3

Pinar, M., Cruciani, C., Giove, S., & Sostero, M. (2014). Constructing the FEEM sustainability index: A Choquet integral application. *Ecological Indicators, 39*, 189–202. https://doi.org/10.1016/j.ecolind.2013.12.012

Pingali, P., & Plavšić, M. (2022). Hunger and environmental goals for Asia: Synergies and trade-offs among the SDGs. *Environmental Challenges, 7*, 100491. https://doi.org/10.1016/j.envc.2022.100491

Pradhan, P., Costa, L., Rybski, D., Lucht, W., & Kropp, J. P. (2017). A systematic study of sustainable development goal (SDG) interactions. *Earth's Future, 5*(11), 1169–1179. https://doi.org/10.1002/2017ef000632

Renaud, F. G., Zhou, X., Bosher, L., Barrett, B., & Huang, S. (2022). Synergies and trade-offs between sustainable development goals and targets: Innovative approaches and new perspectives. *Sustainability Science, 17*(4), 1317–1322. https://doi.org/10.1007/s11625-022-01209-9

Sachs, J., Schmidt-Traub, G., Kroll, C., Lafortune, G., & Fuller, G. (2018). SDG index and dashboards report 2018. Bertelsmann Stifung and Sustainable Development Solutions Network (SDSN).

Saiu, V., Blečić, I., & Meloni, I. (2022). Making sustainability development goals (SDGs) operational at suburban level: Potentials and limitations of neighbourhood sustainability assessment tools. *Environmental Impact Assessment Review, 96*, 106845. https://doi.org/10.1016/j.eiar.2022.106845

Sam, L. (2021). Sustainable agriculture for a sustainable food system: A discussion of synergies and trade-offs within the SDGs framework. *Regions*. https://doi.org/10.1080/13673882.2021.00001089

Schmidt-Traub, G., Kroll, C., Teksoz, K., Durand-Delacre, D., & Sachs, J. D. (2017). National baselines for the sustainable development goals assessed in the SDG index and dashboards. *Nature Geoscience, 10*(8), 547–555. https://doi.org/10.1038/ngeo2985

Sen, A. K. (1999). *Development as freedom*. Oxford University Press.

Shannon, C. E. (1948). A mathematical theory of communication. *Bell System Technical Journal, 27*(3), 379–423. https://doi.org/10.1002/j.1538-7305.1948.tb01338.x

Shapley, L. S. (1953). A value for n-person games. In H. W. Kuhn & A. W. Tucker (Eds.), *Contributions to the theory of games*. Princeton University Press.

Shukla, S., Mohanty, B., & Kumar, A. (2018). Strategizing sustainability in e-commerce channels for additive manufacturing using value-focused thinking and fuzzy cognitive maps. *Industrial Management & Data Systems, 118*(2), 390–411. https://doi.org/10.1108/imds-03-2017-0122

Singh, P. K., & Chudasama, H. (2021). Pathways for climate resilient development: Human well-being within a safe and just space in the 21st century. *Global Environmental Change, 68*, 102277. https://doi.org/10.1016/j.gloenvcha.2021.102277

Spaiser, V., Ranganathan, S., Swain, R. B., & Sumpter, D. J. T. (2016). The sustainable development oxymoron: Quantifying and modelling the incompatibility of sustainable development goals. *International Journal of Sustainable Development & World Ecology, 24*(6), 457–470. https://doi.org/10.1080/13504509.2016.1235624

Stern, S., Harmacek, J., Krylova, P., & Htitich, M. (2022). *2022 Social progress index methodology summary*. Social Progress Imperative (In Press). www.socialprogress.org/global/methodology

Stiglitz, J. E., Fitoussi, J. P., & Durand, M. (2019). *For Good Measure*.

Streeten, P. (1994). Human development: Means and ends. *The American Economic Review, 84*(2), 232–237.

Sugeno, M. (1985). An introductory survey of fuzzy control. *Information Sciences, 36*(1–2), 59–83. https://doi.org/10.1016/0020-0255(85)90026-X

Sun, X., Lu, Z., Li, F., & Crittenden, J. C. (2018). Analyzing spatio-temporal changes and trade-offs to support the supply of multiple ecosystem services in Beijing, China. *Ecological Indicators, 94*, 117–129. https://doi.org/10.1016/j.ecolind.2018.06.049

Thomas, A., & Deakin, E. (2017). Managing partnerships for sustainable development: The Berkeley—China sustainable transportation program. *Case Studies on Transport Policy, 5*(1), 45–54. https://doi.org/10.1016/j.cstp.2016.08.005

Torlockyj, N. (1975). *Improvements in the quality of life*. National Planning Association.

Tosun, J., & Leininger, J. (2017). Governing the interlinkages between the sustainable development goals: Approaches to attain policy integration. *Global Challenges, 1*(9), 1700036. https://doi.org/10.1002/gch2.201700036

United Nations (UN). (2018). *The sustainable development goals report 2018*. United Nations. https://unstats.un.org/sdgs/files/report/2018/TheSustainableDevelopmentGoalsReport2018.pdf

United Nations (UN). (2019). *UN global SDG indicators database*. https://unstats.un.org/sdgs/indicators/database/

Valcárcel-Aguiar, B., Murias, P., & Rodríguez-González, D. (2018). Sustainable urban liveability: A practical proposal based on a composite indicator. *Sustainability, 11*(1), 86. https://doi.org/10.3390/su11010086

Weber, H., & Weber, M. (2020). When means of implementation meet ecological modernization theory: A critical frame for thinking about the sustainable development goals initiative. *World Development, 136*, 105129. https://doi.org/10.1016/j.worlddev.2020.105129

Weitz, N., Carlsen, H., Nilsson, M., & Skånberg, K. (2017). Towards systemic and contextual priority setting for implementing the 2030 agenda. *Sustainability Science, 13*(2), 531–548. https://doi.org/10.1007/s11625-017-0470-0

Wong, R., & Heijden, J. (2019). Avoidance of conflicts and trade-offs: A challenge for the policy integration of the United Nations sustainable development goals. *Sustainable Development, 27*(5), 838–845. https://doi.org/10.1002/sd.1944

Yu, T. H. K., & Huarng, K. H. (2024). Causal analysis of SDG achievements. *Technological Forecasting and Social Change, 198*, 122977. https://doi.org/10.1016/j.techfore.2023.122977

Zhu, J., Zhai, Y., Feng, S., Tan, Y., & Wei, W. (2022). Trade-offs and synergies among air-pollution-related SDGs as well as interactions between air-pollution-related SDGs and other SDGs. *Journal of Cleaner Production, 331*, 129890. https://doi.org/10.1016/j.jclepro.2021.129890

Chapter 7
Underlying the Impact of Economic, Social, and Governance Adaptation on Poverty Reduction Under the Shadow of External Financial Inflows—Panel Estimation from the Sub-Saharan Region

Orhan Şanli, Priyanka Sahu, and Laeeq Razzak Janjua

Abstract Social sustainability contributes to sustainable development in multiple ways. Providing justice in society encourages the distribution of economic resources equally. Currently, more than 700 million people live in extreme poverty today and are struggling to fulfil the most basic needs like health education and access to water and sanitation. Due to the pandemic crises, global poverty increased by 8%, and on the other hand, pandemic crises pushed over 70 million into extreme poverty. This chapter explores the impact of financial inflow on poverty in the Sub-Saharan Africa region under the shadow of adaptation of economic, social, and governance readiness. By using a panel of 38 Sub-Saharan African countries over the period of 2000–2022, applying fixed effect Driscoll-Kraay standard error estimation, and also a two-step system generalized method of moments estimation, the findings reveal that economic prosperity, foreign direct investment, economic and social readiness decrease poverty in sample countries. Similar results are also affirmed by using fixed effect panel quantile regression estimation.

Keywords FDI · Remittance · Aid · Financial inflow · Poverty · Economic growth

O. Şanli
Adnan Menderes Üniversitesi-Aydin, Aydin, Turkey
e-mail: orhan.sanli@adu.edu.tr

P. Sahu
Ghani Khan Choudhary Institute of Engineering and Technology, Maligram, West Bengal, India
e-mail: Priyanka@gkciet.ac.in

L. R. Janjua (✉)
WSB University, Dąbrowa Górnicza, Poland
e-mail: laeeq.janjua@wsb.edu.pl

M. S. Ben Ali and E. Lechman (eds.), *Sustainable Economic Development*,
https://doi.org/10.1007/978-981-97-3767-3_7

7.1 Introduction

Sub-Saharan Africa's persistent poverty despite increasing global integration and rising financial inflows warrants policy attention and adapted reforms oriented toward equitable growth. With over 40% of people in Sub-Saharan Africa living on under $1.90 per day as of 2015, the region contends with disproportionate income poverty transcending low averages (Beegle & Christiaensen, 2019). Furthermore, human development indicators include health, education, and quality of life trail global standards. Yet recent decades have brought rising foreign direct investment, aid, and remittances into African countries, posing potential for poverty alleviation amidst accusations of limited localized benefits (Ajayi & Ndikumana, 2014). Hence, investigating channels linking external finances—entailing adapted policy usage—to inclusive growth and poverty impacts can unlock reform opportunities.

Notable poverty differentials between fast-growing African nations utilizing external finances reveal complex below-the-surface transformations affecting distributional outcomes (Fosu, 2015). The "growth elasticity of poverty" quantifying responsiveness of poverty rates to per capita income growth demonstrates wide variations driven by initial inequality levels alongside differential sectoral poverty exposures during structural change (Christiaensen et al., 2002; Loayza & Raddatz, 2010). External resource flows reshape economic structures, demonstrating poverty effects conditioned on supportive domestic institutions and policies facilitating broad-based participation (Balchin et al., 2019). Governance capacities determining institutional quality, policy coherence, regional coordination, corporate regulation, and export orientation moderate external finances' growth and employment linkages (Anyanwu, 2012; Asongu & le Roux 2017). Hence, personalized analysis of economic, institutional, and social dynamics underpinning uneven gains is imperative.

Financial globalization's poverty impacts also significantly vary based on differing external account exposures—whether inflows from remittances, aid, FDI, or outflows through illicit financing (Kwenda & Ntuli, 2020; Munemo, 2022). Remittance-receiving households demonstrate direct income and consumption growth (Gupta et al., 2009). FDI-linked jobs and production chains foster multiplier impacts, conditional on factor market access (Gohou & Soumaré 2012). Official flows boost public spending and investment, dependent on efficacy (Arndt et al., 2015). Yet outflows drain domestic resources needed for poverty programmes (Adegbite & Sayek, 2011). Adapted usage across flows hinges on absorption capacities determined by the existing administrative apparatus (Wamboye et al., 2013). Measurements, hence, necessitate nuanced approaches accounting for composition over solely aggregate volume effects.

Moreover, disaggregating national poverty estimates into sectoral, rural–urban, gender, and subnational dynamics reveals particularly marginalized groups left behind amidst uneven gains induced through globalization (Beegle & Christiaensen, 2019). Rural smallholders confront export competition under liberalization despite facing intrinsic productivity barriers, warranting policy protections (Akpan et al.,

2013). Youth encounter constrained mobility pathways to industrial wage jobs from external investment (Anyanwu, 2013). Impacts, therefore, demonstrate demographic and spatial heterogeneity unlikely captured in national averages. Identifying intersectional groups facing compounded deprivations is vital (Klasen & Lamanna, 2009). Thus, analysing Sub-Saharan Africa's poverty response to rising but unevenly distributed global financial integration necessitates assessing economic restructuring impacts on marginalized demographic groups under prevailing distributional regimes.

7.2 Literature Review

Research reveals a complex interplay of economic, social, and governance factors driving poverty outcomes in Sub-Saharan Africa. While financial sector liberalization policies accelerate growth, Dandume (2014) determined their poverty impacts depend on distributional changes and supporting institutions rather than automatic translation. Analysing foreign aid's significant poverty reduction capacity, Mahembe (2018, 2021) found good governance and effective targeting strategies optimize outcomes, confirming the importance of accountable institutional quality. Assessing financial inflows, Kwabena Twerefou et al. (2020) showed remittances and foreign direct investment boost growth whereas official development assistance reduces growth due to recipient country governance weaknesses, highlighting absorption capacitates' pivotal role. Thus, multiple analyses indicate both growth-stimulating reforms and equity-focused redistribution mechanisms under accountable governments are essential for translating macro-level policy gains into tangible poverty reductions across Sub-Saharan African societies.

Foreign direct investment (FDI) constitutes a major financial inflow for many Sub-Saharan African countries and has emerged as an important driver of growth and poverty alleviation. Analysing data from 31 countries over 2009–2019, Ayenew (2022) found FDI has a robust, significant positive impact on economic growth. At the regional level, Gaston (2009) used a human development and FDI perspective to confirm a notable positive relationship between FDI and poverty reduction in Central and East Africa, though impacts vary across African regions. These FDI-induced gains likely result from positive production externalities and technological diffusion to domestic firms. In contrast, evaluating aid flows over nearly three decades, Edmore and Odhiambo (2021) reveal that total official development assistance (ODA) significantly reduces poverty by financing pro-poor expenditures. Though when examining economic growth, Daniel et al. (2020) find ODA negatively and significantly affects Sub-Saharan African growth, attributed to inefficient allocation and difficulties translating ODA into capital investments due to poor institutional quality. Thus, while ODA alleviates poverty directly through transfers, inefficient utilization prevents it from enhancing productive capacity and growth.

Given increasing remittance significance, Gupta et al. (2009) assessed their developmental role in Sub-Saharan Africa and determined remittances have an independent, poverty-reducing impact, explained by direct income transfers to rural households. This aligns with Ayenew's (2022) statistically insignificant finding. Evaluating inequality impacts, Thamaga and Ncanywa (2021) demonstrate remittances strongly reduce income inequality over the long term across SADC countries. All studies recommend policies to lower transfer costs and encourage productive remittance usage through financial access and investment options to magnify development impacts.

Yet surpassing these inflows are Africa's massive illicit financial outflows, estimated between \$854 billion and \$1.8 trillion from 1970 to 2008 per Nkurunziza (2012), who highlights the association between high illicit outflows and poverty. Dandume and Malarvizhi (2014) analysed financial liberalization's impact on poverty and determined that while it accelerates growth, distributional changes and supporting policies are essential for translating gains into poverty reduction. This finding reveals that both growth levels and existing inequalities influence poverty outcomes.

In assessing the effectiveness of foreign aid, which comprises sizable budget share in many Sub-Saharan states, Deborah and Knack (2004) note that high aid maintained over long periods risks government dependence and institutional damage, reiterating the need for strong institutions and policy environments outlined by Daniel et al. (2020). Analysing drivers of poverty reductions, Sillah (2012) demonstrates per capita GDP growth that has the largest poverty-reducing effect in Sub-Saharan Africa while non-economic measures may fail to trickle down provoking debate on best pro-poor policies.

In light of growth differentials between countries utilizing foreign capital, Chimere O. et al. (2017) employed regional estimation techniques to assess short- and long-run impacts across Sub-Saharan Africa. Results revealed integration lags likely explain the insignificant short-run effects of financial inflows on growth, while long-run variation suggests nuanced inflow-growth linkages. Recommendations include differentiation between inflow policies and simultaneous adoption of openness strategies. Context-specific analysis is imperative as Gaston (2009) established, finding FDI's poverty alleviation role varies across African regions.

While the literature focused extensively on economic growth linkages, Letsoalo and Ncanywa (2021) uniquely analysed how four key foreign flows affect persistent income inequality across Southern African Development Community members over 1981–2014. Employing cointegration and causality techniques, the study confirms remittances and foreign aid reduces inequality in the long term whereas FDI and lending increase inequality. Causality tests also detect bidirectional relationships, underscoring complex associations between external finance and inequality outcomes. As such, channelling different flows into pro-poor, inequality-conscious investments could profoundly shape distributional outcomes.

On the fiscal outcome front, Prakash and Cabezon's (2008) work marked the first Sub-Saharan study verifying that enhancing public financial management quality significantly improves overall fiscal balance and lowers external debt levels in African low-income countries. Thus, domestic governance capacities inform debt sustainability and economic stability. In a comparative study on remittances in Sub-Saharan African and Latin American and Caribbean countries, Adenutsi and Christian (2010) employed an augmented Human Development Index to illustrate remittances' role differs across regions—visible positive impacts in the African context contrasting with detrimental development effects in Latin America and the Caribbean. Thus, relying extensively on remittances remains questionable for both regions.

In assessing fiscal policy options, Kimaro et al. (2017) analysed government expenditure efficiency's impact across 25 Sub-Saharan countries and concluded that while increasing government spending accelerates growth, interacting expenditure with efficiency does not heighten the growth effect. Hence rather than emphasis on efficiency, government expenditure prioritization requires cost–benefit analysis regarding growth linkages in African low-income economies.

Considering social protection policies, Hodges et al. (2013) used quantitative simulations to compare cash transfers' effectiveness for alleviating monetary poverty against advancing broader human development in an oil-producing middle-income African country (Republic of Congo) relative to a low-income country (Côte d'Ivoire). Although results showed sizable poverty reduction capacity in both countries, cash transfers demonstrated weaker human development impacts, challenging feasibility for national rollout—an important lesson for the limits of certain poverty interventions' financing and political backing in the African context.

In assessing simultaneous adoption of economic reform measures, Sakyi (2011) employed cointegration techniques to demonstrate that while increased trade and financial openness boost Africa's long-run incomes, expanding democracy negatively drags income—reflecting inappropriate institutional structures currently undermining democracy's development dividends. Thus, ensuring democratic quality and economic justice will be key for Africa to harness globalization's benefits while mitigating adverse distributional consequences.

Extant empirical evidence reveals enduring and disproportionate income poverty across Sub-Saharan Africa despite rising integration into global financial markets and capital flows in recent decades (Beegle & Christiaensen, 2019). While aggregate foreign direct investments, development assistance, and remittances have grown, localized translational impacts remain contested, necessitating analysis of binding constraints (Ajayi & Ndikumana, 2014).

Heterogeneous poverty outcomes between externally financed, fast-growing African economies point to multifaceted below-the-surface transformations shaping distributional consequences, including sectoral poverty exposures and inequality traps (Fosu, 2015). Macro-econometric analyses highlight the significant influence of structural economic shifts triggered by globalization on the responsiveness of national poverty rates to per capita income growth trajectories or "growth elasticities" (Christiaensen et al., 2002; Loayza & Raddatz, 2010). The literature underscores endogenous institutional capacities, policy coherence, and regulatory environments as critical mediators determining whether external resource flows catalyze broad-based participation and poverty alleviation (Anyanwu, 2012; Asongu & Le Roux, 2017).

Similarly, financial globalization's heterogeneous poverty effects contingent on composition underscores the need for disaggregated assessments by source and directionality—whether illicit outflows, remittance inflows, aid, or foreign direct investment (Munemo, 2014; Kwenda & Ntuli, 2020). Poverty impacts fundamentally rely on absorption capacities across government, private sector, and household units (Wamboye et al., 2013).

Structural shifts also demonstrate demographic and spatial heterogeneities with disproportionate rural, informal sector, and female poverty entrenchment attributed to uneven mobility barriers, productivity traps, and human capital barriers limiting upward convergence (Beegle & Christiaensen, 2019). Identifying multiplicatively marginalized groups is critical (Klasen & Lamanna, 2009). The studies by Ben et al. (2020) reveal the multifaceted impact of corruption on economic and financial systems. They highlight in their research work, how corruption escalates the likelihood of banking crises, particularly affecting low-income countries by inducing excessive risk-taking in banks rather than merely affecting profitability. This relationship exhibits a nonlinear pattern with two distinct regimes, indicating a corruption threshold that drives regime shifts. Moreover, their 2019 research on MENA countries demonstrates the varying contributions of capital accumulation and technological progress to economic growth, with human capital playing a significant role in specific contexts (Ben Ali & Acikgoz, 2019). Swaleheen et al. (2019) further elaborate on corruption's effects, noting its nonlinear influence on public spending across 134 countries, wherein corruption positively impacts health spending but negatively affects education expenditure. This finding is supported by Saha and Ben (2017), who highlight the contrasting impact of corruption on health and education spending, emphasizing the need for targeted policy interventions. Additionally, Ben Ali and Sassi (2016) illustrate a positive relationship between corruption measures and inflation rates, particularly pronounced in developing and emerging nations. Finally, Pradhan et al. (2022) underscore how corruption in less developed countries diverts crucial government funds from essential sectors, advocating for the role of information technology (ICT) in managing resources and combating corruption. Their study reveals the interconnectedness of ICT infrastructure, corruption control, and government spending, emphasizing the pivotal role of strong ICT infrastructure and strategic government spending in mitigating corruption and associated risks.

In conclusion, while foreign finance enables poverty alleviation and growth under the right conditions, policy and institutional constraints alongside sustainable financing challenges temper development progress in Sub-Saharan Africa. Per capita income growth remains imperative for poverty reduction, though inequality issues warrant distribution-conscious reforms ensuring macro gains reach the marginalized. More comparative, context-specific analyses evaluating channels linking financial flows to poverty, inequality and human welfare are critical to elucidating nuanced, actionable policy insights for the region.

7.3 Economics, Social, and Governance Readiness

A nation's preparedness in these areas frequently determines its capacity to navigate challenges, foster growth, and sustain progress. Thus, economic, social, and governance readiness are crucial pillars that support the stability, development, and resilience of nations in a rapidly evolving global landscape. Let us delve into each dimension to understand their significance and interplay.

Economic Readiness: This relates to a nation's ability to generate new ideas, adjust to changes, and prosper in the worldwide economy. It encompasses elements like commercial activities, infrastructure construction, financial strategies, and technical progress. An economically prepared nation has robust institutions facilitating investment, innovation, and corporate expansion. Furthermore, it is also referred as economic assets, such as road, railways and airports and all other assets which are directly or indirectly under the control of government (Moteff et al., 2004). An adequate infrastructure, including energy, digital, transportation, and communication, is the fundamental basis for economic progress. Investments in these sectors enhance overall economic efficiency, productivity, and interconnectedness. Economic diversification refers to expanding economic activity in producing and distributing commodities and services. While it may not directly result in increased output, broadening the financial foundation of countries can promote their stabilization. Economic diversification should be examined through the lens of sustainable development to guarantee the enduring stability of the economy (Anyaehie & Areji, 2015). Countries that place a high priority on R&D and technological developments have an advantage in the international marketplace. Innovation promotes economic growth, productivity gains, and advancement. A strong financial system is essential for economic preparedness because it promotes investment, entrepreneurship, and financial inclusion for all societal groups. It also guarantees stability and credit availability.

The economic preparedness of a nation is vital for its prosperity. Sub-Saharan countries must prioritize the development of economic preparedness in order to effectively eradicate poverty (UNDP, 2019). Preparedness encompasses innovation, infrastructure development, financial strategy, economic diversification, and the adoption of new technology. Innovation and research play a vital role in enabling Sub-Saharan nations to tackle local challenges in agriculture, education, and healthcare, thereby alleviating poverty. According to the UNDP (2019), innovation plays a vital role in attaining the Sustainable Development Goals, with the primary focus being on reducing poverty. Research and development yield advancements in agricultural production through the use of advanced inputs such as drought-resistant seeds. These advancements offer the potential for significantly increased output to farmers in the area who have low productivity. The implementation and evaluation of remote learning, telemedicine, and data technologies facilitate the provision of high-quality education and healthcare services to remote rural communities (World Bank, 2018). A dynamic innovation ecosystem that fosters the development of cutting-edge technology has the potential to significantly enhance human progress and alleviate poverty.

Infrastructure is crucial for fostering economic growth and alleviating poverty. According to the World Bank (2017), enhancing infrastructure in Sub-Saharan Africa has the potential to increase productivity by 40% and foster fair and balanced development. Constructing reliable roadways, bridges, railway systems, seaports, and airports enhances both domestic and international trade. The African Development Bank is providing financial support for more than 50 significant transport corridors that link landlocked countries to coastal gateways through the Infrastructure Consortium for Africa (AfDB, 2021). Enhanced product, service, and human mobility facilitates economic interactions and market entry, leading to increased income and decreased poverty. According to Power Africa (2021), the widespread adoption of decentralized renewable energy systems enhances the productivity of rural smallholder farmers and informal microenterprises. Access to electricity amplifies opportunities in underprivileged areas by stimulating poverty-alleviating catalysts. Investing in infrastructure enhances efficiency, collaboration, and innovation to support comprehensive and sustainable development.

Impoverished individuals and communities require financial tools to allocate resources towards enhancing their human capital and engaging in economic endeavours. According to Demirgüç-Kunt et al. (2018), there is evidence to suggest that a greater level of financial system depth is associated with a decrease in income inequality and poverty across 150 countries. Credit unions, mobile money, and microfinance in Sub-Saharan Africa can facilitate the adoption of new technology and the establishment of informal enterprises among farmers. Financial inclusion enables individuals to save money as a precautionary measure against unexpected crises that could lead to significant deprivation. Sufficient insurance safeguards valuable possessions, enabling families to undertake enterprising ventures that gradually lift them out of poverty. Financial systems that are inclusive facilitate self-enhancement.

Restricted export bases and excessive price volatility can lead to external shocks that undermine poverty reduction efforts in many Sub-Saharan African economies. Economic diversification serves as a safeguard against economic downturns. According to the World Bank, Cameroon's shift towards agricultural exports shielded it from recession during the oil price crises of the 1970s and 80s, while Nigeria, which heavily relies on oil, experienced a recession (Gill et al., 2014). The expansion of export sectors such as fashion, horticulture, and meat processing leads to the creation of numerous employment opportunities including those that require low levels of skill. This is particularly important in order to accommodate the increasing number of young individuals in the workforce (AfDB, 2016). The augmentation and constancy of the financial productivity derived from multiple channels diminishes both national and individual impoverishment.

The fields of digital, biological, and materials sciences offer significant potential for accelerating poverty reduction. A study conducted in rural Kenya discovered that the availability of mobile phone coverage facilitated the connection between farmers and fishermen with improved marketplaces, resulting in a 15% increase in their revenue (Aker & Mbiti, 2010). Enhanced productivity and reduced crop or catch wastage facilitated poverty alleviation. Ahead are a few poverty-alleviating concepts, such as gene drives for malaria prevention, 3D printed houses, and distributed ledger systems for financial inclusion. Sub-Saharan nations can benefit from continuous advancements by giving priority to advanced technical readiness and assimilation.

Therefore, in order to achieve sustainable economic growth, it is necessary to deliberately cultivate interconnected elements of innovation, infrastructure, finance, diversity, and technology that are specifically designed to meet the local requirements. This expansion eradicates poverty in Sub-Saharan Africa through long-term and equitable prosperity. The economic preparedness pillars serve as the driving force for eradicating poverty (Demirgüç-Kunt et al., 2018; Aker & Mbiti, 2010). It is feasible to eradicate extreme poverty within a single generation through a steadfast dedication to expedite progress in every aspect, working together in harmony.

Social Readiness: This refers to a country's ability to guarantee the health, equality, and cohesiveness of its citizens. It includes social safety nets, healthcare, education, and cultural diversity. An adept and well-informed labour force fosters economic expansion and ingenuity. Access to a reasonably priced, superior education equips individuals with the necessary skills to make a substantial economic impact. Human development has a significant impact on economic progress. Every component of human progress, in particular, is expected to have a different effect on economic growth. For example, education significantly impacts labour productivity; furthermore, labour productivity positively impacts economic growth in the long run.

Social preparedness, which includes investments in human capital and social safety nets, is an essential component of poverty alleviation efforts in Sub-Saharan Africa. Considering that more than 40% of the population in Sub-Saharan region lives in poverty, it is very necessary to construct resilience by implementing health, education, and social protection systems that are specifically adapted to the area. Ethiopia, Rwanda, and Ghana are three countries that provide examples that illustrate the potential for social preparedness to reduce poverty.

Improving healthcare preparedness by investing in community health professionals, clinics, and insurance accessibility has far-reaching implications on poverty reduction. To begin, this improvement may be accomplished via investments. Nsanzimana et al. (2020) report that Rwanda's community-based health insurance programme covers more over 90% of the population, making it possible for individuals to have cheap access to medical care and providing protection against financial risk. Achieving 90% immunization coverage and a 50% decline in under-five mortality from 2000 to 2015 is a remarkable achievement that paves the road for people to escape poverty. This is accomplished by combining this with the establishment of resilient health systems and the decentralization of primary care.

Education, which is the most reliable means of climbing out of poverty, is fuelled by the correct nourishment that is made possible by health. It is noteworthy that Ethiopia has made significant progress in the field of education, as seen by the fact that the primary net enrolment ratio has increased from 20% in 1990 to 86% by 2019 (UNDP Africa, 2021). These achievements were driven by centralized planning and funding that prioritized school building, teacher training, and curriculum development in alignment with antipoverty aims; nonetheless, it is still necessary to improve quality. The combination of education and health care is a significant investment that helps the disadvantaged and prepares human capital.

In addition, the growth of social safety net programmes on a large scale, which includes cash transfers, school food, and public works, has been an important factor in the reduction of poverty. For instance, since 2008, the Livelihood Empowerment Against Poverty (LEAP) Programme in Ghana has been providing unrestricted cash, enrolment in healthcare programmes, and vocational assistance to hundreds of thousands of families who are living in severe poverty (UNDP Africa, 2021). In the course of rigorous impact assessments, it was discovered that LEAP was able to cut poverty headcounts by 11%, demonstrating the capacity of social protection to elevate those who are vulnerable (UNICEF, 2021).

The development of digital identification systems, payment platforms, and social registries is an essential component of these initiatives, as it enhances the delivery of social programmes to the region's financially disadvantaged population. According to Bold et al.'s research from 2020, using digitized targeting and integrating transfer receipts across paperless systems helps reduce leakage while simultaneously enhancing efficiency.

Generally speaking, making health, education, social protection, and digital acceleration, the basic pillars of social preparedness may yield significant poverty reduction for disadvantaged people living in Sub-Saharan Africa. These pro-poor social investments, when combined with supportive governance and economic policies, provide tried and true formulas for eradicating poverty in the midst of the demographics of destiny (Beegle & Christiaensen, 2019).

Governance readiness: An effectively managed nation creates an environment conducive to economic advancement, social unity, and stability. Preparedness for effective governance: The pillars of governance preparation include efficient institutions, transparent policies, adherence to the rule of law, and accountability. A society benefits from having strong legal frameworks and ensuring adherence to the rule of law, since this promotes predictability, stability, and transparency. Clear and explicit regulations foster trust among the general public and investors. Transparent governance methods cultivate trust between the government and its stakeholders. Efficient governance requires the implementation of accountability protocols and robust measures to mitigate corruption. Countries with robust political systems are more capable of implementing enduring programmes and attracting investments. Political stability is a crucial prerequisite for achieving long-term and sustained economic expansion. Inclusive governance involves actively including individuals in decision-making processes to ensure that policies accurately reflect the needs and objectives of the population.

Effective governance is essential for achieving long-term poverty alleviation in Sub-Saharan region (Sulaiman et al., 2020). Countries may successfully implement policies that benefit the poor and promote fair economic development by developing competent institutions, upholding the rule of law, promoting accountability, and empowering local communities. Rwanda, Ghana, and Kenya have implemented governance innovations that demonstrate effective strategies for decreasing poverty and preparing for good governance.

After the occurrence of a genocide Rwanda has shown exceptional political stability and implemented effective governance reforms since 2000. Rwanda implemented economy-wide reforms such as public sector capacity development, strict anti-corruption measures, and decentralized service delivery. As a result, Rwanda achieved the second highest ranking in Africa on transparency and competitiveness indices (Mwai & Gisore, 2022). These improvements in governance, together with smart economic reforms and social investments, resulted in a reduction in monetary poverty by 20 percentage points between 2006 and 2020 (NISR, 2022).

Furthermore, Ghana has solidified its position as an exemplary African democracy by successfully undergoing three peaceful political changes since the year 2000. Implementing institutional mechanisms to maintain checks and balances between the executive and legislative branches guarantees accountability and the formulation of policies that prioritize the needs of the people (Agyeman-Duah, 2014). As a result, Ghana achieved lower-middle income status by 2011 and had successive decreases in poverty rates.

Additionally, the 2010 Kenyan constitution enshrined unprecedented devolution of powers to 47 county governments who receive billions for locally-led development. Although there may be difficulties in putting it into practice, this system of localized autonomy and proximal governance allows for the development of customized anti-poverty programmes that address the specific requirements of different communities (D'Arcy & Cornell, 2016). As an example, the agricultural investment and safety net programmes implemented in Makueni County resulted in a 50% reduction in local poverty rates over a period of 10 years (O'Brien & Hanauer, 2022).

Implementing e-governance technology strengthens the capacity and responsibility of the government, which is crucial for efficient provision of public services to impoverished people. The process of converting property registries, tax payments, and procurement into digital formats simplifies operations and decreases instances of fraud, hence allowing for the allocation of resources towards poverty programmes (World Bank, 2016).

The effectiveness of poverty alleviation policies in the area is heavily dependent on the overall improvement of governance preparedness in terms of public sector performance, decentralization, and technical capabilities (Sulaiman et al., 2020; Mwai & Gisore, 2022). By implementing effective governance practices, Sub-Saharan nations have the potential to accomplish long-term poverty reduction via economic growth and improved social services at the local level.

7.4 Research Framework and Empirical Research

This research work questions the growth-finance-poverty trilemma by presenting empirical evidence from Sub-Saharan countries under the shadow of ESG readiness. The growth variable is percentage change in economic growth per year (GDP) extract from World Bank development indicators; whereas the external financial inflow based on foreign direct investment (FDI), remittance inflow (REM), and official development assistance (ODA) along with Trade (T) as percentage of GDP. These variables also extracted from World Bank development indicators. Furthermore, the economic readiness, social readiness, and governance readiness are extracted from University of Notre Dame Global Adaptation Initiative, ND-GAIN Country Index database. Similarly, the Per capita consumption expenditure growth-Proxy for poverty is extracted from PovcalNet (World Bank development indicators). The research framework is mentioned in Fig. 7.1.

Similarly, the summary statistics thus mean mode, medium, and number of observations are mentioned in Table 7.4. LN refers to the log form of used variables; thus increasing the elasticity among the variables. Considering the fact that Sub-Saharan countries are diverse in nature to different economic structures; therefore, numerous indicators consist of outliers (see Figs. 7.2, 7.3, 7.4, 7.5, 7.6, 7.7, 7.8, 7.9, and 7.10). Furthermore, the matrix graph is mentioned in Fig. 7.11.

7.4.1 *Empirical Methodology*

The study uses static and dynamic models to achieve the research objectives systematically. These estimation approaches are suitable and used by similar studies (Fosu,

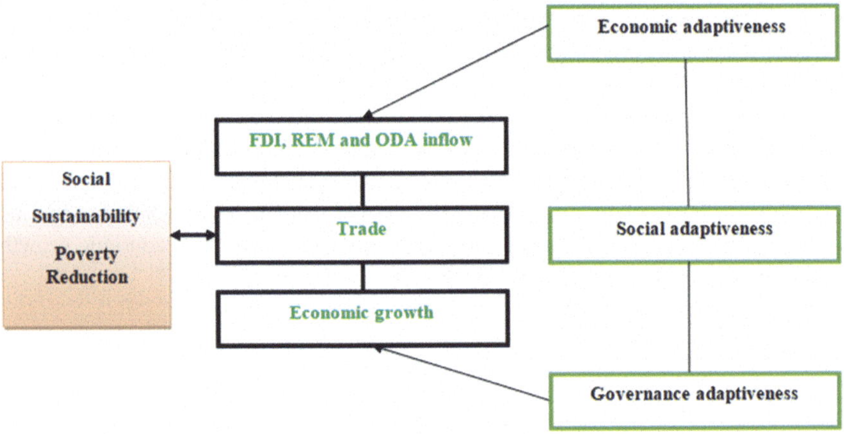

Fig. 7.1 Research framework

2017; Marrero & Serv'en, 2018; Adeleye et al., 2020) in addition to using a short panel data of 38 countries (N) across 23 years (T), hence, $N > T$. Similarly, the adoption of these techniques serves as robustness for one another in order to observe the consistency of the relation among the variables of interest. This research uses static (Eixed Effect Driscoll-Karay estimation) as main estimation and dynamic model (system-GMM) as robust estimation models to systematically achieve the research objective along with panel quantile estimation as robust estimation.

7.4.1.1 Fixed Effect Driscoll-Karay (Main Estimation)

By considering the possibility of heterogeneity, I estimate the results by applying the fixed effect model. It incorporates the sampled country's specific policies and practices of the transport model and shows the effects in the intercept coefficient. "α_{1j}". The intercept of one country differs from the other country but is time-invariant. The following equation (A) captures the countries' specific effects by takings the different economic, geological and social characteristics. The Driscoll and Kraay (1998) robust-standard errors-type approach, which accounts for heteroskedasticity, autocorrelation, and cross-sectional dependence, is used in this study. The empirical strategy on which Driscoll and Kraay (1998) estimation is based on that, it takes averages of the product between independent variables and the residuals, and then to use these values in a weighted HAC estimator to produce standard errors that are now have the added feature of being robust against cross-sectional dependence (Hoechle, 2007; Vogelsang, 2012). Ramoutar (2017) also affirms in his empirical analysis that, the advantage of fixed effects with Driscoll and Kraay standard errors is that the problems of heteroscedasticity, autocorrelation, and cross-sectional dependence are all corrected. Similar estimation methodology is also used by Arogundade et al. (2022), Thompson (2011), Topcu and Gulal (2020), and Rehman and Ahmad (2022).

$$Y_{jt} = \alpha_{1j} + \beta_1 X_{1jt} + \beta_2 X_{2jt} + \beta_3 X_{Njt} + \beta_i \sum_{i=1}^{i=N} L_{ji} + u_{jt} \tag{7.1}$$

7.4.1.2 System-GMM (Robust Estimation)

The panel data usually violate the autocorrelation and homoscedastic OLS assumptions. So, there are the chances of serial correlation and heteroskedasticity problems that may unbiased the estimates. The disturbance term correlates to any variable in the model causing the serial correlation problem (Attari et al., 2016). The variations in the variance of error terms across the observation cause the issue of heteroskedasticity (Simpson, 2012). Auto-correlation and heteroskedasticity problems can be resolved by applying the GMM models proposed by Arellano and Bond (1991).

Usually, heteroskedasticity, autocorrelation, and heterogeneity are the potential problems in the dynamic panel data sets. These problems are solved by applying the panel GMM (Attari et al., 2016). GMM is suitable for the panel data with more significant cross-section identifiers than periods (Blundell & Bond 1998). The following Eq. 7.2 captures the dynamic effects of the model to test the hypotheses.

$$Y_{jt} = \alpha_0 + (1 - \beta_1)Y_{jt-1} + \beta_2 X_{1jt} + \beta_3 X_{2jt} + \beta_4 X_{Njt} + \beta_i \sum_{i=1}^{i=N} L_{ji} + v_t + \varepsilon_{i,t} \quad (7.2)$$

where $v_t = u_t - (1 - \beta_1)u_{t-1}$

Similarly, Table 7.1 reports the impact of impact of economic growth and other exogenous variables under the shadow of ESG readiness indicator on poverty according to the Fixed effect Driscoll-Kraay Estimation for Sub-Saharan countries. Column (1) shows that among eight exogenous variables trade, foreign direct investment and economic growth appear to have positive impact on poverty reduction in sample countries. 1 unit increase in trade, foreign direct investment and economic growth causes 0.166%, 0.555%, and 0.368% of poverty reduction by increasing per capita expenditure. Similarly, in Columns (2), (3), and (4) by incorporating ESG readiness along with other five exogenous variables separately, the estimated results indicate consistent signs of trade, foreign direct investment and economic growth; however, the magnitude of coefficients is slightly different. Political, economic, and institutional index appeared to be insignificant across all the estimations. Column (5) presents the impact of five exogenous variables along with ESG readiness on poverty reduction. The estimated results are consistent with previous estimations as mentioned in column (1)–(4). According to the results in all five estimation mentioned in Column (1) to Column (5), an averagely 1 unit increase in trade decreases poverty by 0.153%, respectively, in the presence of 0.065% of foreign direct investment inflow. Similarly; 1 unit increase in economic growth causes 0.375% decrease in poverty by increase per capita expenditure. Furthermore, the estimated results also reveal that unit increase in economic and social readiness tends to decrease poverty by 0.401% and 3.977%, respectively.

Table 7.2 reports the impact of economic growth and other exogenous variables under the shadow of ESG readiness indicator on poverty according to the system-GMM estimation which is a robust estimation. The results indicate that the lagged dependent variable coefficient lag.LNPOV is significant and positive for all estimated equations, as indicated in columns (1) to (5). The lag value has a positive impact on the poverty, which means that LNPOV has had a positive and significant impact on the current value of PCE growth rate in the past few years. Column (1) shows among six exogenous variables foreign direct investment and economic growth appear to have positive impact on poverty reduction in given sample of Sub-Saharan countries. According to the system-GMM estimation, 1 unit increase in foreign direct investment and economic growth reduces poverty by 0.068% and 0.310%, respectively. Similarly, in Column (2), (3), and (4) by incorporating ESG readiness along with other five exogenous variables, the estimated results indicate consistent signs of foreign

Table 7.1 Fixed effect—DK estimation

Variable	Model 1	Model 2	Model 3	Model 4	Model 5
LNGDP	0.368 (0.151)**	0.388 (0.152)**	0.335 (0.174)**	0.375 (0.155)**	0.336 (0.149)**
LNT	0.166 (0.830)*	0.178 (0.083)*	0.111 (0.084)*	0.181 (0.093)*	0.072 (0.096)*
LNFDI	0.055 (0.738)**	0.075 (0.859)**	0.078 (0.077)**	0.065 (0.455)**	0.079 (0.076)***
LNREM	−0.022 (0.042)	0.082 (0.471)	0.014 (0.045)	0.023 (0.042)	0.015 (0.046)
LNODA	0.005 (0.038)	0.057 (0.039)	0.048	0.002 (0.038)	0.004 (0.035)
ECO	–	0.811 (0.53)**	–	–	0.401 (0.476)*
SOC	–		4.030 (0.073)***	–	3.977 (0.812)***
GOC	–	–	–	0.172 (0.390)	0.244 (0.375)
Constant	0.537 (0.400)	0.589 (0.582)	0.588 (0.589)	0.546 (0.414)	0.576 (0.516)

direct investment and economic, which are statistically significant; however, the magnitude of coefficients is slightly different. Column (5) presents the impact of five exogenous variables along with ESG readiness on poverty reduction for Sub-Saharan region. The estimated results reveal that foreign direct investment and economic growth rate positively reduce poverty. According to the results in (5), 1 unit increase in foreign direct investment and economic growth in Sub-Saharan countries causes 0.084% and 0.306% of poverty reduction. Similarly, the estimated results also reveal that ESG readiness in Sub-Saharan countries causes an increase in poverty reduction; thus 1 unit increase in ESG readiness decreases poverty by 1.214%, 3.608%, and 1.553%, respectively, by increase per capita consumption expenditure. In system-GMM, post-estimation diagnosis shows that AR (2) and Hansen test, the probability values are insignificant, which implies that there are no correlations, and similarly; instruments are reliable, consistent, and suitable for drawing inferences. Compared to the results of fixed effect Driscoll-Kraay Estimation (main analysis) which are mentioned in the previous table with the estimated results system-GMM, the results' interpretation will emphasize the consistency of the sign of the coefficient and their statistical significance if any. Therefore; the results indicate that economic growth, foreign direct investment, economic readiness, and social readiness decrease poverty. The coefficients are positive and have statistical significance across all the entire models. Similarly, according to Table 7.3, the panel quantile regression, the economic growth reduces poverty from low to high all quantiles thus at 20th, 40th, 60th, and 80th. Similar results were also derived for trade and foreign direct investment in the presence of economic and social readiness. However, the governance readiness also reduces poverty at 60th and 80th quantiles, thus, coefficient appears to be positive and significant.

Table 7.2 System GMM

Variable	Model 1	Model 2	Model 3	Model 4	Model 5
Lag.POV	0.048 (0.062)***	0.049 (0.067)***	0.032 (0.064)***	0.045 (0.061)***	0.029 (0.062)***
GDP	0.310 (0.184)*	0.313 (0.181*	0.291 (0.195)*	0.317 (0.185)*	0.306 (0.192)*
T	0.140 (0.237)	0.152 (0.232)	0.099 (0.226)	0.079 (0.239)	0.086 (0.214)
FDI	0.068 (0.072)*	0.677 (0.085)*	0.077 (0.072)*	0.070 (0.071)*	0.084 (0.077)*
REM	0.045 (0.035)	0.048 (0.034)	0.036 (0.032)	0.049 (0.039)	0.052 (0.058)
ODA	0.031 (0.033)	0.039 (0.037)	0.058 (0.055)	0.016 (0.053)	0.057 (0.058)
ECO		0.319 (0.056)**	–	–	1.214 (0.775)**
SOC		–	2.872 (1.552)*	–	3.608 (1.665)*
GOC		–	–	0.807 (0.587)**	1.553 (0.778)**
Constant	−0.286 (0.944)	−0.234 (0.981)	−0.533 (1.010)	−0.402 (0.963)	−0.739 (1.036)
Number of groups	38	38	38	38	38
Number of instruments	8	8	8	8	11

7.5 Discussion

This research work questions the growth-finance-poverty trilemma by presenting empirical evidence from Sub-Saharan countries under the shadow of ESG readiness, which fill the lacuna in the literature. Likewise, this empirical work highlighted the essence of ESG readiness to achieve social sustainability by reducing poverty in sample countries. The estimated results reveal that the positive and significant relations depict that cumulative effect in FDI inflow improved per capita consumption expenditure, thus reducing poverty in Sub-Saharan countries. The relationship between FDI inflow and per capita expenditure is because of ongoing economic development in Sub-Saharan countries. FDI inflow increases the amount of exiting capital in the host country, leading to a rise in the marginal physical product of labour and will increase wages, directly impacting the individual consumption expenditure. FDI inflow creates more jobs and increase individual wages, thus increasing their purchasing power ability and consumption expenditure. These findings align with previous studies such as Chishti et al. (2021) and Wangdi and Wangdi (2021), which also found a similar relationship between FDI inflow and per capita consumption expenditure. Likewise, several other studies such as Sharma and Gani (2004),

Table 7.3 Quantile regression

Variable	Quantiles							
	20		40		60		80	
	Coefficient	Standard error	Coefficient	Standard error	Coefficient	Standard error	Coefficient	Standard error
LNGDP	0.543*	0.282	0.371*	0.200	0.225*	0.140	0.184*	0.130
LNT	0.049***	0.244	0.111***	0.167	0.136**	0.206	0.229**	0.142
LNFDI	0.055**	0.088	0.128**	0.048	0.074**	0.061	0.098**	0.040
LNREM	0.020	0.077	0.016	0.036	0.005**	0.032	0.014	0.037
LNODA	0.023	0.080	0.040	0.066	0.044	0.068	0.021	0.059
LECO	0.546**	1.311	0.108**	0.834	0.029**	0.867	0.987**	0.668
SOC	3.638**	1.879	3.812**	1.134	5.105**	0.932	5.493**	0.669
GOV	1.903	1.336	1.149	0.577	0.416**	0.878	1.488**	0.461
_cons	−0.867	1.402	0.837	0.732	1.201	0.829	1.896	0.635

Table 7.4 Summary statistics

Variable	Obs	Mean	Standard deviation	Min	Max
LNPOV	539	0.913996	1.224213	−6.37076	4.134686
LNGDP	774	1.398585	0.822456	−3.50656	4.149148
LNT	874	4.104764	0.464763	0.99282	5.169335
LNFDI	813	0.75301	1.233286	−4.60517	4.164803
LNREM	873	−0.04467	1.965516	−8.58389	3.985764
LNODA	872	1.380446	1.33126	−6.03684	4.104455
ECO	874	0.304555	0.112526	0	0.819906
SOC	874	0.198893	0.055308	0.078926	0.359757
GOV	874	0.394438	0.118358	0.160324	0.697117

Calvo and Hernandez (2006), Nunnenkamp et al. (2007), Israel (2014), Fowowe and Shuaibu (2014) also indicate that FDI inflow positively reduces poverty in different regions which reinforces the findings of the study. This research also suggests that economic growth increases per capita consumption expenditure, thus reducing poverty in Sub-Saharan countries. The results explain that economic growth will enhance the consumption capacity of an individual to meet his needs, reducing poverty. Previous studies affirm the outcome of this finding for different regions and countries (Fosu, 2017; Garza-Rodriguez, 2018; Adeleye et al., 2020). Furthermore, the findings also suggest that the economic readiness and social readiness also increase per capita consumption expenditure thus decreases poverty; it is due to the fact economic and social readiness increases the job opportunities, which increases the well-being of the people in Sub-Saharan region.

7.6 Conclusion and Policy Implication

Considering the fact that poverty is a global issue and a lot of countries in Sub-Saharan countries face stagnant growth. Policymakers must prioritize Sub-Saharan Africa and enact customized reforms to achieve equitable growth since the area continues to experience persistent poverty despite its increasing integration into the global economy and rising financial inflows. This study empirically analyses a nexus between external inflow, poverty, and growth in Sub-Saharan countries under the shadow of governance, social, and economic readiness (ESG) factors. By analysing data from 38 Sub-Saharan African nations between 2000 and 2022, and employing a two-step Generalized Method of Moments estimate along with fixed effect Driscoll-Karray standard error, the findings indicate that economic development, foreign direct investment (FDI), social and economic preparation have a mitigating effect on poverty in the examined countries. Based on the discussion, this government should adopt the mentioned below policies in order to reduce poverty in Sub-Saharan

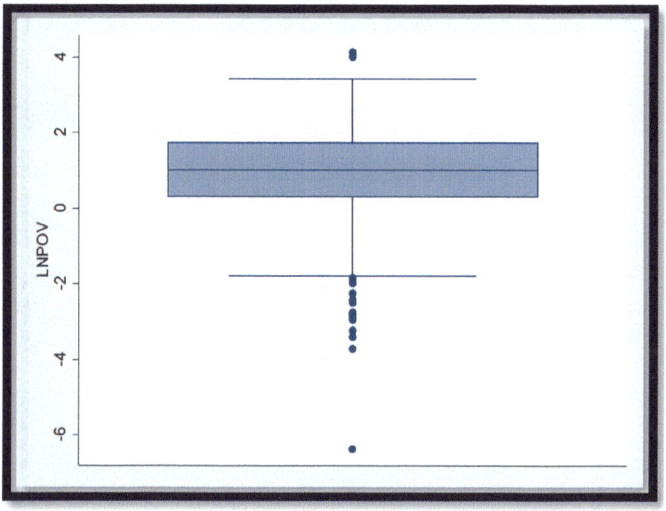

Fig. 7.2 Box plot poverty

Region. The government should provide income support to people below the poverty line (such as the Ehsas income support programme in Pakistan). Government should increase the progressive tax as it improves the purchasing power capability. Government should provide essential health and education services or facilities without any cost. By using AID in infrastructure directly improves employment and increases individual income. Government should create employment opportunities. Government should encourage the use of renewable energy resources (such as solar panels) to reduce the cost of living and improve purchasing capacity. The government should encourage improving agriculture productivity, as it impacts the employment sector of industrial and agriculture sectors. Saving schemes based on tax networks could also be helpful in poverty reduction, such as the different rates for tax filers versus non-tax filers.

Appendix

See Table 7.4.

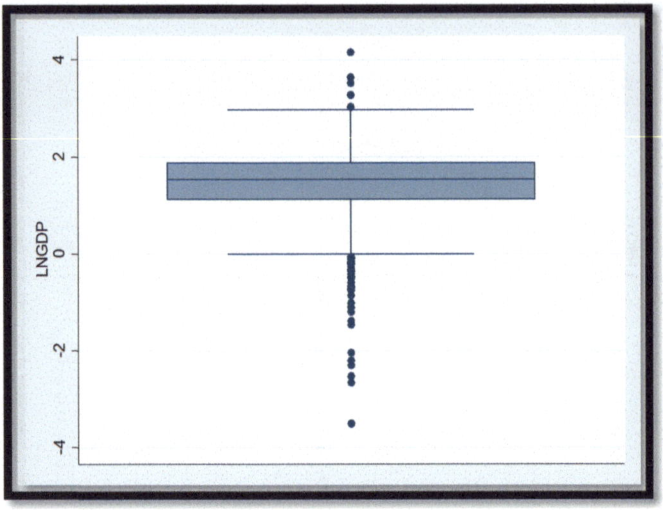

Fig. 7.3 Box plot GDP

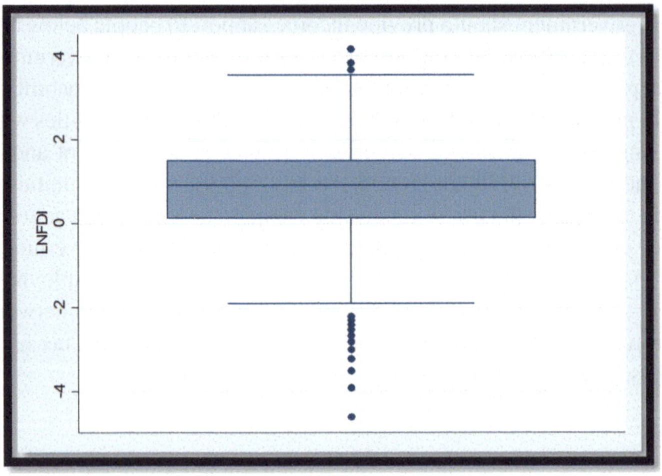

Fig. 7.4 Box plot FDI

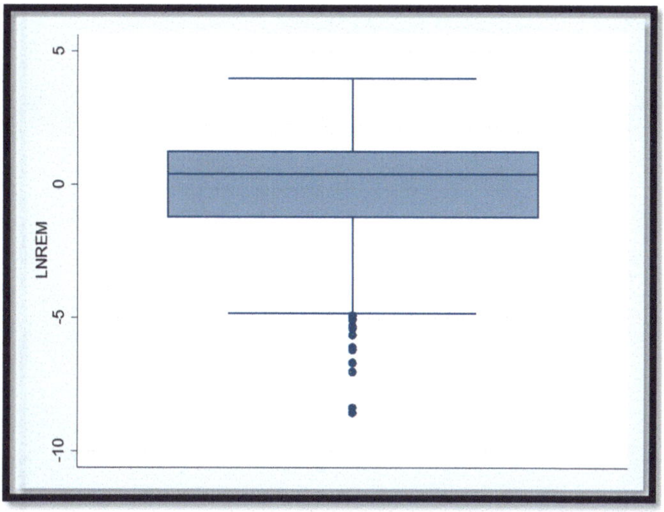

Fig. 7.5 Box plot Rem

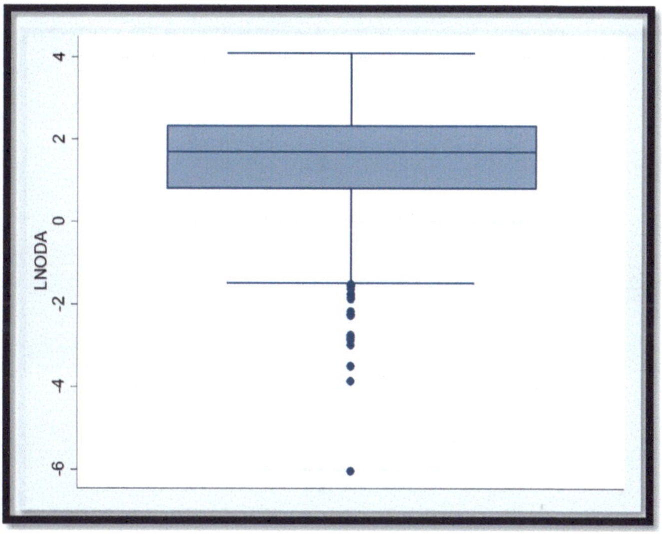

Fig. 7.6 Box plot ODA

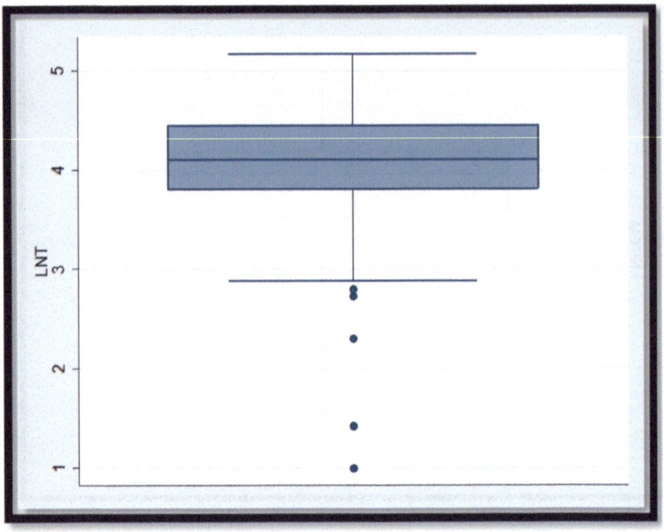

Fig. 7.7 Box plot T

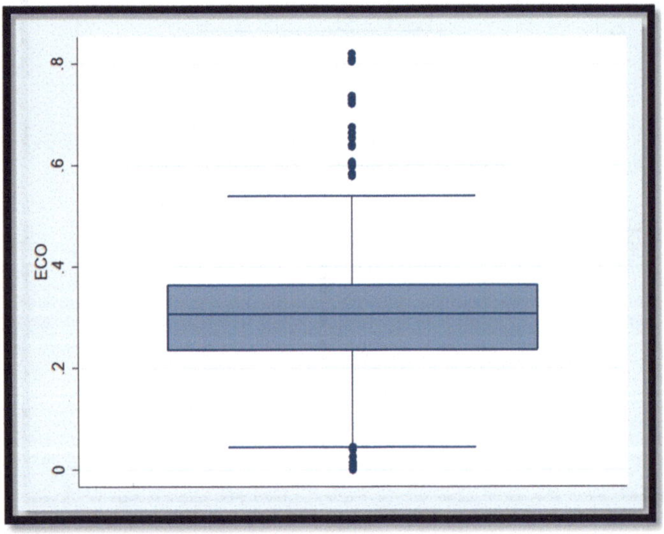

Fig. 7.8 Box plot economic readiness

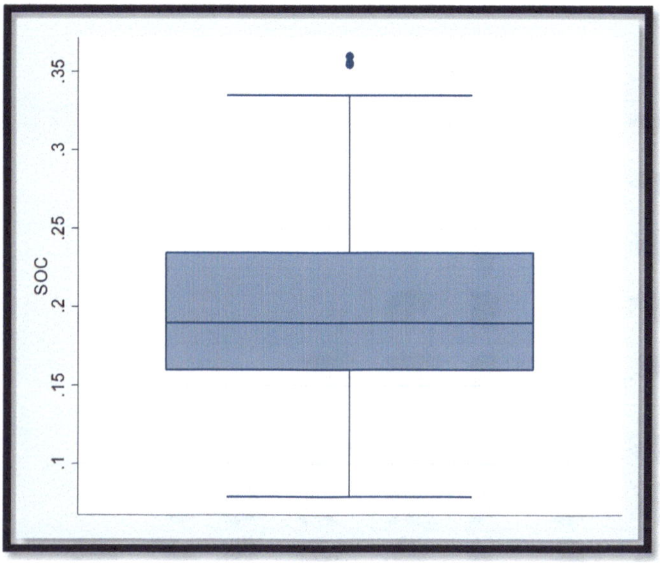

Fig. 7.9 Box social readiness

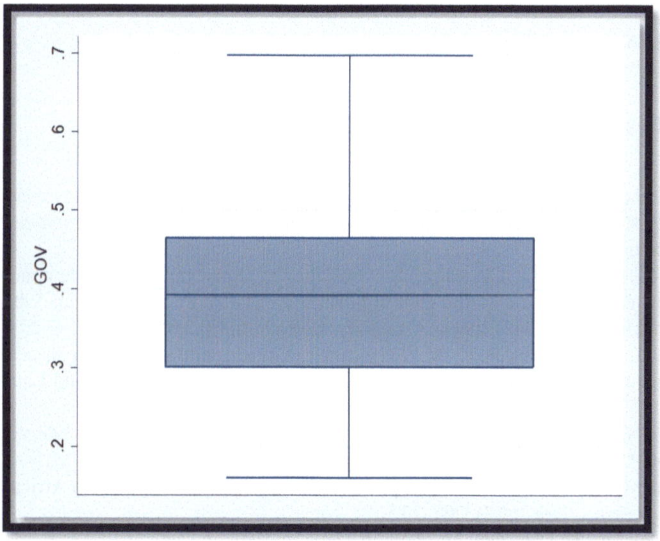

Fig. 7.10 Box plot governance readiness

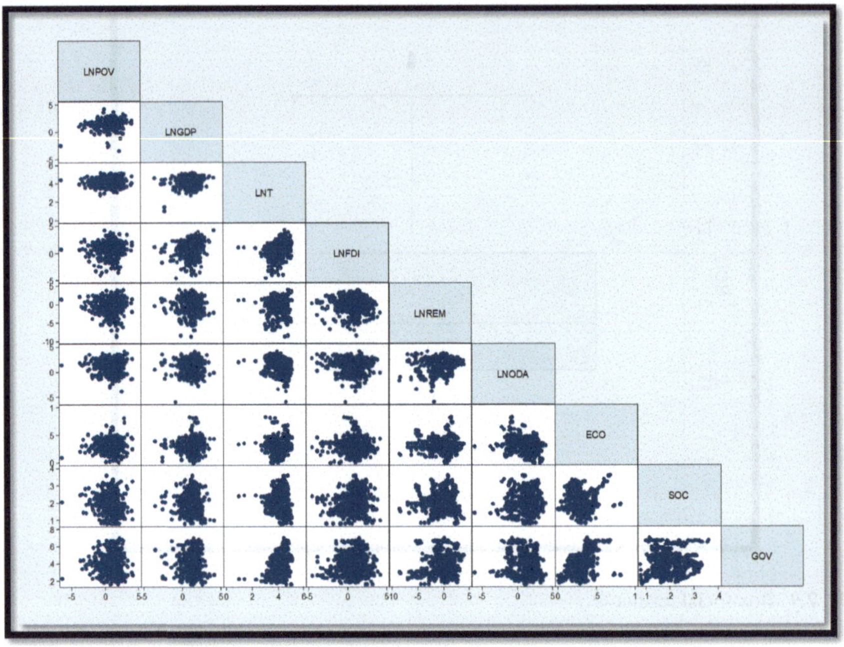

Fig. 7.11 Matrix graphs

References

Adegbite, E. O., & Ayadi, F. S. (2011). The role of foreign direct investment in economic development: A study of Nigeria. *World Journal of Entrepreneurship, Management and Sustainable Development, 6*(1/2), 133–147.

Adeleye, B. N., Gershon, O., Ogundipe, A., Owolabi, O., Ogunrinola, I., & Adediran, O. (2020). Comparative investigation of the growth-poverty-inequality trilemma in Sub-Saharan Africa and Latin American and Caribbean Countries. *Heliyon, 6*(12), e05631.

Adenutsi, D. E., & Christian, R. K. A. (2010). International remittances–The panacea for underdevelopment? A comparative panel data analysis of Sub-Saharan Africa and Latin America, 235–247.

Ajayi, S. I., & Ndikumana, L. (Eds.). (2014). *Capital flight from Africa: Causes, effects, and policy issues*. OUP Oxford.

Aker, J. C., & Mbiti, I. M. (2010). Mobile phones and economic development in Africa. *Journal of Economic Perspectives, 24*(3), 207–232.

Akpan, B., et al. (2013). Determinants of credit access and demand among poultry farmers in Akwa Ibom State, Nigeria. *American Journal of Experimental Agriculture*, 3.2, 293.

Anyanwu, J. C. (2012). Why does foreign direct investment go where it goes? New evidence from African countries. *Annals of Economics and Finance, 13*(2), 425–462.

Anyanwu, J. C. (2013). Characteristics and macroeconomic determinants of youth employment in Africa. *African Development Review, 25*(2), 107–129.

Arellano, M., & Bond, S. (1991). Some tests of specification for panel data: Monte Carlo evidence and an application to employment equations. *The Review of Economic Studies, 58*(2), 277–297.

Arndt, C., Jones, S., & Tarp, F. (2015). Assessing foreign aid's long-run contribution to growth and development. *World Development, 69*, 6–18.

Arogundade, S., Mduduzi, B., & Hassan, A. S. (2022). Spatial impact of foreign direct investment on ecological footprint in Africa. *Environmental Science and Pollution Research, 29*(34), 51589–51608.

Asongu, S. A., & Le Roux, S. (2017). Enhancing ICT for inclusive human development in Sub-Saharan Africa. *Technological Forecasting and Social Change, 118*, 44–54.

Attari, M. I. J., Hussain, M., & Javid, A. Y. (2016). Carbon emissions and industrial growth: An ARDL analysis for Pakistan. *International Journal of Energy Sector Management, 10*(4), 642–658.

Ayenew, B. B. (2022). The impact of foreign financial inflows on the economic growth of sub-Saharan African countries: An empirical approach. *Cogent Economics & Finance, 10*(1), 2123888.

Beegle, K., & Christiaensen, L. (Eds.). (2019). *Accelerating poverty reduction in Africa.* World Bank Publications.

Ben Ali, M. S., & Acikgoz, S. (2019). Where does economic growth in the Middle Eastern and North African countries come from? *Quarterly Review of Economics and Finance, 73*, 172–183. (Applied).

Ben Ali, M. S., Sassi, S. (2016). The Corruption-Inflation Nexus: Evidence from Developed and Developing Countries. *The BE Journal of Macroeconomics, 16*(1), 125–144. (Applied).

Ben Ali, M. S., Fhima, F., & Nouira, R. (2020). How does corruption undermine banking stability? A threshold nonlinear framework. *Journal of Behavioral and Experimental Finance, 27*. (Applied).

Blundell, R., & Bond, S. (1998). Initial conditions and moment restrictions in dynamic panel data models. *Journal of Econometrics, 87*(1), 115–143.

Bräutigam, D. A., & Knack, S. (2004). Foreign aid, institutions, and governance in Sub-Saharan Africa. *Economic Development and Cultural Change, 52*(2), 255–285.

Bukhari, M. S. S. (2012). Governance and income determinants of poverty reduction: A cross-country analysis of Sub-Saharan African Countries. *African Journal of Business Management, 6*(3). Chapter 1–2.pdf.

Calvo, C. C., & Hernandez, M. A. (2006). Foreign direct investment and Poverty in Latin America. *Leverhulme Centre for Research on Globalisation and Economic Policy*, University of Nottingham.

Chishti, M. Z., Ahmad, M., Rehman, A., & Khan, M. K. (2021). Mitigations pathways towards sustainable development: Assessing the influence of fiscal and monetary policies on carbon emissions in BRICS economies. *Journal of Cleaner Production, 292*, 126035.

Christiaensen, L. J., Demery, L., & Paternostro, S. (2002). *Growth, distribution and poverty in Africa: Messages from the 1990s* (Vol. 614). World Bank Publications.

Dandume, M. Y., & Malarvizhi, A.C. (2014). Does financial liberalization, spur economic growth and poverty reduction in six Sub-Saharan African countries; panel unit root and panel vector error correction tests.

Driscoll, J. C., & Kraay, A. C. (1998). Consistent covariance matrix estimation with spatially dependent panel data. *Review of Economics and Statistics, 80*(4), 549–560.

Fosu, A. K. (2015). Growth, inequality and poverty in Sub-Saharan Africa: Recent progress in a global context. *Oxford Development Studies, 43*(1), 44–59.

Fosu, A. K. (2017). Growth, inequality, and poverty reduction in developing countries: Recent global evidence. *Research in Economics, 71*(2), 306–336.

Fowowe, B., & Shuaibu, M. I. (2014). Is foreign direct investment good for the poor? New evidence from African countries. *Economic Change and Restructuring, 47*(4), 321–339.

Garza-Rodriguez, J. (2018). Poverty and economic growth in Mexico. *Social Sciences, 7*(10), 183.

Gohou, G., & Soumaré, I. (2012). Does foreign direct investment reduce poverty in Africa and are there regional differences? *World Development, 40*(1), 75–95.

Gohou, G. (2009). Impact of Fdi on poverty reduction in Africa: Are there regional differences?

Gupta, S., Pattillo, C. A., & Wagh, S. (2009). Effect of remittances on poverty and financial development in Sub-Saharan Africa. *World Development, 37*(1), 104–115.

Hodges, A., et al. (2013). Are cash transfers a realistic policy tool for poverty reduction in Sub-Saharan Africa? Evidence from Congo-Brazzaville and Côte d'Ivoire'. *Global Social Policy, 13*(2), 168–92.

Hoechle, D. (2007). Robust standard errors for panel regressions with cross-sectional dependence. *The Stata Journal, 7*(3), 281–312.

Iheonu, C. O., et al. (2017). A Pooled Mean Group Estimation of Capital Inflow and Growth in sub–Saharan Africa. *The Romanian Economic Journal*, 105–121.

Israel, A. O. (2014). Impact of foreign direct investment on poverty reduction in Nigeria. *Journal of Economics and Sustainable Development, 5*(20), 34–45.

Kimaro, E. L., Keong, C. C., & Sea, L. L. (2017). Government expenditure, efficiency and economic growth: a panel analysis of Sub Saharan African low income countries. *African Journal of Economic Review*, 5(2), 34–54.

Kwabena Twerefou, D., et al. (2020). Do foreign financial inflows impact on economic growth? Evidence from sub-Saharan Africa. *Studies of Applied Economics, 38*(2).

Letsoalo, T. E., & Ncanywa, T. (2021). Effects of external financial flows on income inequality in selected southern African development community member states. *Africa's Public Service Delivery and Performance Review, 9*(1).

Loayza, N. V., & Raddatz, C. (2010). The composition of growth matters for poverty alleviation. *Journal of Development Economics, 93*(1), 137–151.

Mahembe, E., & Odhiambo, N. M. (2021). Does foreign aid reduce poverty? A dynamic panel data analysis for Sub-Saharan African Countries'. *The Journal of Economic Inequality, 19*(4), 875–93.

Mahembe, E. (2018). Does foreign aid reduce poverty in Sub-Saharan Africa? A dynamic panel-data analysis. *Research Papers in Economics.*

Munemo, J. (2022). Do African resource rents promote rent-seeking at the expense of entrepreneurship? *Small Business Economics, 58*(3), 1647–1660.

Nkurunziza, J. D. (2012). Illicit financial flows: A constraint on poverty reduction in Africa illicit financial flows and Africa's Slow Poverty Reduction.

Nunnenkamp, P., Schweickert, R., & Wiebelt, M. (2007). Distributional effects of FDI: How the interaction of FDI and economic policy affects poor households in Bolivia. *Development Policy Review, 25*(4), 429–450.

Pradhan, P. R., Nair, S. M., Arvin B. M., Ben Ali, M. S. (2022). *Intertemporal linkages between government expenditure and corruption in a digital economy in digitalization and economic development: Insights from developing countries.* Ben Ali, M. S. (Eds). Routledge, Taylor and Francis.

Prakash, T., & Cabezon, E. R. (2008). Public financial management and fiscal outcomes in Sub-Saharan African heavily-indebted poor countries. *Political Economy: Budget.*

Ramoutar, E. (2017). *The investment cost of currency crises in emerging markets: An empirical treatment from 1994–2015.* Duke University, Durham, North Carolina.

Rehman, F. U., & Ahmad, E. (2022). The effect of climate patterns on rice productivity in Pakistan: an application of Driscoll and Kraay estimator. *Environmental Science and Pollution Research, 29*(35), 53076–53087.

Saha, S., & Ben Ali, M. S. (2017). Corruption and economic development: New evidence from the middle Eastern and North African Countries. *Economic Analysis and Policy, 54*, 83–95. (Applied).

Sakyi, D. (2011). Economic globalisation, democracy and income in Sub-Saharan Africa: A panel cointegration analysis.

Simpson, D. (2012). Knowledge resources as a mediator of the relationship between recycling pressures and environmental performance. *Journal of Cleaner Production, 22*(1), 32–41.

Swaleheen, M., Ben Ali, M. S., & Temimi, A. (2019). Corruption and public spending on education and Health. *Applied Economics Letters, 26*(4), 321–325. (Applied).

Thompson, S. B. (2011). Simple formulas for standard errors that cluster by both firm and time. *Journal of Financial Economics, 99*(1), 1–10.

Topcu, M., & Gulal, O. S. (2020). The impact of COVID-19 on emerging stock markets. *Finance Research Letters, 36*, 101691.

Vogelsang, T. J. (2012). Heteroskedasticity, autocorrelation, and spatial correlation robust inference in linear panel models with fixed-effects. *Journal of Econometrics, 166*(2), 303–319.

Wamboye, E., Adekola, A., & Sergi, B. S. (2013). Economic growth and the role of foreign aid in selected African countries. *Development, 56*(2), 155–171.

Wangdi, D., & Wangdi, P. (2021). A Study on Impact of Foreign Direct Investment (FDI) on Poverty Reduction in Bhutan. *Sherub Dome, 14*(14).

The manufacturer's authorised representative in the EU is Springer Nature Customer Service Centre GmbH, Europaplatz 3, 69115 Heidelberg, Germany. If you have any concerns regarding our products, please contact ProductSafety@springernature.com

Printed and bound by CPI Group (UK) Ltd, Croydon, CR0 4YY
01/10/2026
02225862-0018